Acta Anatomica

International Journal of Anatomy, Embryology and Cell Biology

Founded 1945 by R. Chambers, G. Glimstedt, T. Peterfi and G. Wolf-Heidegger
Continued 1962–1974 by E.A. Boyden, 1955–1980 by A. Delmas,
1972–1980 by F. Walberg, 1945–1980 by G. Wolf-Heidegger,
1981–1988 by R. O'Rahilly, Davis, Calif., 1989–1990 by G.E. Goslow, Jr., Providence, R.I.,
1981–1992 by W. Lierse, Hamburg

KARGER

Printed in Switzerland
on acid-free paper by
Reinhardt Druck, Basel

Appears monthly:
3 volumes per year
(12 issues)

Guidelines for Authors

Acta Anatomica

Aims and Scope

Acta Anatomica is an international forum for communication among scientists interested in morphology at all levels of organization, from subcellular to macroscopy, and in morphogenesis, with emphasis on humans and higher vertebrates. The journal offers rapid exchange of novel and significant information gained from experimental and theoretical investigations on all aspects that are pertinent to the formation and transformation of biological structure.

Topics of special interest include

Vertebrate embryology (including developmental and reproductive biology), cell biology, neuroanatomy, functional and clinical anatomy (including application of modern techniques of imaging). Papers highlighting the links between structural and functional aspects are preferred.

Types of Papers

The journal publishes: (1) Original papers on experimental research. (2) Theoretical contributions. These will appear in the section 'Hypotheses and Concepts' and may include provocative hypotheses, provided that they are innovative, well founded and show promise for new avenues of research. (3) Short reviews/perspectives. While (1) and (2) are submitted contributions, reviews/perspectives (3) are invited. Suggestions concerning reviews on current topics of high interest are welcome.

All articles submitted must be original. All manuscripts must be approved by all authors, and this has to be certified by the first author with his signature. Manuscripts may not have been published previously either in whole or in part, except in abstract form, and may not be under consideration by any other journal. Reported studies using animal and human subjects must have been approved by the appropriate institutional review boards.

Submission

Manuscripts from North and South America should be sent to:

Prof. A.W. English, PhD
Emory University School of Medicine
Dept. of Anatomy and Cell Biology
Atlanta, GA 30322 (USA)
Telefax +1 404 727–3677
E-Mail: acta@cellbio.emory.edu

Manuscripts from Japan and South-East Asia to:

Prof. Dr. Osamu Ohtani
Department of Anatomy II
Toyama Medical and Pharmaceutical University
2630 Sugitani
Toyama, 930-01 (Japan)
Telefax +81 764 34 5010

Manuscripts from all other countries to:

Prof. Dr. med. Dr. rer. nat. H.-W. Denker
c/o S. Karger AG
Editorial Dept. 'Acta Anatomica'
P.O. Box
CH–4009 Basel (Switzerland)
Telefax +41 61 306 14 34

All manuscripts are subject to peer review. Submission of an article for publication implies the transfer of the copyright from the author to the publisher upon acceptance. Accepted papers become the permanent property of Acta Anatomica and may not be reproduced by any means, in whole or in part, without the written consent of the publisher. It is the author's responsibility to obtain permission to reproduce illustrations, tables, etc. from other publications.

Format and Arrangement

Manuscripts should be written in English and be as concise as possible. They should be submitted in quadruplicate (with four sets of original illustrations), typewritten double-spaced on one side of the paper, with a wide margin. Lines should be numbered on the margin.

Title page: Title (main title underlined), the authors' names, and the institute where the work was conducted, a short title for use as running head, the full address of the author to whom correspondence should be sent. Please also supply phone and fax numbers, as well as your e-mail address. If different, indicate in addition a full address for revisions and proof reading separately.

Key words: For indexing purposes provide a list of up to 5 key words.

Abstract: Each paper needs an abstract of up to about 20 lines summarizing the essential new informations communicated.

Subheadings: Introduction; Materials and Methods; Results; Discussion.

Small type: Paragraphs which can or must be set in smaller type (methods, etc.) should be indicated with a 'p' (petit) in the margin on the left-hand side.

Footnotes: Abbreviations used in the paper will be printed as a footnote on the first page. Please list them on a separate page of the manuscript. Otherwise, footnotes should be avoided. When essential, they are numbered consecutively and typed at the foot of the appropriate page.

Nomenclature: The nomenclature used should conform, wherever possible, to the current edition of the Nomina Anatomica, Nomina Anatomica Veterinaria, or to the Nomina Anatomica Avium.

Tables and illustrations: Tables and illustrations (both numbered in Arabic numerals) should be prepared on separate sheets. Tables require a heading and figures a legend, also prepared on a separate sheet. For the reproduction of illustrations, only good drawings and original photographs can be accepted; negatives or photocopies cannot be used. Due to technical reasons, figures with a screen background should not be submitted. When possible, group several illustrations on one block for reproduction (max. size 181 × 223 mm) or provide crop marks. On the back of each illustration, indicate its number, the author's name, and 'top' with a soft pencil.

Color illustrations: During the review process, the decision will be made as to which color illustrations of a manuscript can be reproduced free of charge. Authors will be informed of this at the time of acceptance of their paper. Otherwise, these illustrations will be reproduced at the author's expense. Up to 6 color illustrations per page can be integrated into the text at the reduced price of CHF 660.– / USD 545.00 per page.

References: References in the text should be made by author's name followed by the year of publication, arranged chronologically, then alphabetically. When there are more than two authors, use the first author's name and et al. When references are made to more than one paper by the same author, published in the same year, designate them as a, b, c, etc. The list of references should include only those publications which are cited in the text. Material submitted for publication but not yet accepted may be cited in the text as 'unpublished data' but must not be included in the reference list. In the list of references, arrange references alphabetically listing all authors, then year of publication. Abbreviate journal names according to Index Medicus.

Examples

(a) Papers published in periodicals: Sun, J., H. Koto, K.F. Chung (1997) Interaction of ozone and allergen challenges on bronchial responsiveness and inflammation in sensitised guinea pigs. Int Arch Allergy Immunol *112:* 191–195.

Hadler, W.A., S.R. Silveira (1981) Técnica histoquímica que permite discriminar o colesterol de substâncias afins e de seus ésteres. Cienc Cult 33(suppl): 586–587.

(b) Monographs: Matthews DE, Farewell VT: Using and Understanding Medical Statistics, ed 3, revised. Basel, Karger, 1996.

(c) Edited books: Parren, P.W.H.I., D.R. Burton (1997) Antibodies against HIV-1 from phage display libraries: Mapping of an immune response and progress towards antiviral immunotherapy; in Capra JD (ed): Antibody Engineering. Chem Immunol. Basel, Karger, vol 65, pp 18–56.

Cover illustrations

The Journal Acta Anatomica will publish cover illustrations relevant to any area of modern anatomical research. Cover illustrations will be chosen primarily on the basis of aesthetic appeal and not necessarily from articles in that issue of the journal. Authors or readers are encouraged to submit illustrations suitable for the cover. Only black-and-white illustrations can be accepted. Please submit two copies (max. 13 × 16 cm) to the appropriate editors. Submissions should include a brief caption on the back of the illustration.

Galley proofs

Unless indicated otherwise, galley proofs are sent to the first-named author and should be returned with the least possible delay. Alterations made in proofs, other than the correction of printer's errors, are charged to the author.

Reprints

Order forms and a price list are sent with the proofs. Orders submitted after the issue is printed are subject to considerably higher prices.

KARGER

Fax + 41 61 306 12 34
E-Mail karger@karger.ch
www.karger.com

The Guidelines for Authors are available at:
www.karger.com/journals/aan/aan_gl.htm

General Information

Acta Anatomica

Publication data: 'Acta Anatomica' is published 12 times annually. Volumes 161–163 with 4 issues each appear in 1998.

Subscription rates: Subscriptions run for a full calendar year. Prices are given per volume.
Personal subscription:
Subscription + postage and handling
CHF 315.– + CHF 15.70 Europe
CHF 315.– + CHF 23.70 Overseas
USD 247.50 + USD 15.70
(Must be in the name of, billed to, and paid by an individual. Order must be marked 'personal subscription'.)
Institutional subscription:
Subscription + postage and handling
CHF 630.– + CHF 19.60 Europe
CHF 630.– + CHF 29.60 Overseas
USD 495.00 + USD 19.60
Airmail postage: CHF 24.–, USD 20.00.
(Extra per volume.)

Microforms: Please direct all enquiries and orders to: University Microfilms Inc.
300 North Zeeb Road, P.O. Box 1346
Ann Arbor, MI 48106-1346 (USA)

Subscription orders: Orders can be placed at agencies, bookstores, directly with the Publisher
S. Karger AG, P.O. Box,
CH–4009 Basel (Switzerland)
or with any of the following distributors:
Deutschland: S. Karger GmbH, Postfach
D–79095 Freiburg
France: Librairie Luginbühl
36, bd de Latour-Maubourg, F–75007 Paris
Great Britain, Ireland: S. Karger Publishers
58 Grove Hill Road
Tunbridge Wells, Kent TN1 1SP
USA: S. Karger Publishers, Inc.
26 West Avon Road, P.O. Box 529
Farmington, CT 06085
India, Bangladesh, Sri Lanka:
S. Karger India Subscription Agency
B–5/132 Safdarjung Enclave
New Delhi-110029
Japan: Katakura Libri, Inc.
Take 1 Bldg, 2nd Floor, 4-5-7 Yushima
Bunkyo-ku, Tokyo 113

Change of address: Both old and new address should be sent to the subscription source.

Single issues and back volumes: Information can be obtained through the Publisher.

Advertising: Correspondence and rate requests should be addressed to the Publisher.

Bibliographic indices: This journal is regularly listed in bibliographic services, including *Current Contents®*.

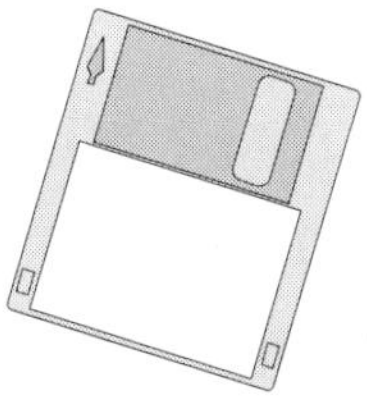

Disk Submission

A **comprehensive version** of these guidelines for the **submission of manuscripts and illustrations on disk** can be accessed online under **www.karger.com/journals/disksub.htm** or requested as a leaflet from the editorial department.

Please follow the general rules on style and arrangement in the **"Guidelines for Authors"** of the journal. Submit a **double-spaced printout** of your manuscript matching the file on disk exactly.

Technical Information

Store only data pertinent to this manuscript. Include your fonts (screen and printer fonts for Macintosh). For a **text only** manuscript, use a 3.5 inch disk. For **text and illustrations,** use 3.5 or 5.25 Magneto Optical Disc; CD-ROM (DOS or hybrid); iomage Zip/DOS (100 MB), or 44 MB, 88 MB or 200 MB SyQuest. Compress extensive data using ZIP format.

Data Description

Label your disk with the **title of the journal or book** for which it is submitted; a **short version** of the paper's title; your **name,** postal and e-mail address, fax and phone number; **hard and software** used, including the version; and the **complete list of the file names.** File names consist of eight characters before the dot identifying the file and three characters after the dot identifying the application.

Text and Format

The **preferred word-processing package is MS-Word.** Other commonly used PC text programs and ASCII formats are accepted.
Update your data when you revise subsequent to the reviewer's or editor's decision and mark revisions clearly. Manuscripts accepted for publication are subject to **copy editing:** changes and/or questions will be marked in the manuscript, which will be returned to you with your disk and the galley proofs for checking.
Enter your text continuously flush left. Do not use any layout functions. Do not split words at the end of a line. **Do not use indentations.** Use a blank line before and a hard return at the end of paragraphs.
Arrange headings flush left. Insert only **one space after punctuation marks.** Insert Greek letters, mathematical symbols etc. with your word-processing program. Use **boldface and italics** as well as sub- and superscript where appropriate.
Tables are part of the text and should be placed at the end of the text file. Separate the individual parts of a table by lines. Use the tabulator instead of spaces. Format operation data on one line in the same column. **Legends to illustrations are part of the text** and should also be listed at the end of the text file.

Illustrations

Store illustrations as separate files: do not embed them in your text. Export **line drawings/vector graphics** as TIF, EPS or WMF format. Use Photoshop for processing and retouching **scanned halftone images.** Save the original scan and the processed version. Export **black and white or color images** as TIF or EPS format in their anticipated size in print.
For detailed information on how to submit illustration data on disk please consult the leaflet or **www.karger.com/journals/disksub.htm.**

KARGER

KI 98 DISK

Fax+ 41 61 306 12 34
E-Mail karger@karger.ch
www.karger.com

The Journal Home Page is available at:
www.karger.com/journals/aan

A contemporary update on human motor control

Perspectives of Motor Behavior and Its Neural Basis

Edited by:
Marie-Claude Hepp-Reymond
Gabriella Marini

This publication provides the reader with a better understanding of some basic principles of motor behavior and gives an update on modern approaches of human motor control. It contains abundant information on the current trends and illustrates the progress from laboratory findings to the investigation of more natural movements as well as of the cognitive aspects of motor behavior. As an additional benefit for the reader, the collected data is put in a historical perspective. Basic and clinical neuroscientists, rehabilitation specialists, physiotherapists and in particular students in system neuroscience, robotics and bioengineering will find this book a noteworthy contribution to the field.

Contents

Perspectives of Motor Behavior and Its Neural Basis
Editors: Hepp-Reymond, M.-C. (Zürich); Marini, G. (Milano)
VIII + 138 p., 60 fig., 2 tab., hard cover, 1997
CHF 98.– / DEM 117.– / USD 85.25
Prices subject to change
DEM price for Germany, USD price for USA only
ISBN 3–8055–6403–1

Fields of Interest: Neurology; Physiology

KI 97431

KARGER

Acta Anatomica

Brainstem (Pre)motor Mechanisms and the Stereotypy of Motor Patterns

Guest Editor
J.L. Dubbeldam, Leiden

25 figures, 2 tables, 1998

KARGER

Basel · Freiburg · Paris · London · New York ·
New Delhi · Bangkok · Singapore · Tokyo · Sydney

KARGER

S. Karger
Medical and Scientific Publishers
Basel · Freiburg · Paris · London
New York · New Delhi · Bangkok
Singapore · Tokyo · Sydney

Drug Dosage
The authors and the publisher have exerted every effort to ensure that drug selection and dosage set forth in this text are in accord with current recommendations and practice at the time of publication. However, in view of ongoing research, changes in government regulations, and the constant flow of information relating to drug therapy and drug reactions, the reader is urged to check the package insert for each drug for any change in indications and dosage and for added warnings and precautions. This is particularly important when the recommended agent is a new and/or infrequently employed drug.

P.O. Box, CH–4009 Basel (Switzerland)
Printed in Switzerland on acid-free paper by
Reinhardt Druck, Basel
ISBN 3–8055–6848–7

Acta Anatomica

Contents Vol. 163, No. 3, 1998

KARGER

Fax+ 41 61 306 12 34
E-Mail karger@karger.ch
www.karger.com

Access to full text and tables of contents, including tentative ones for forthcoming issues: www.karger.com/journals/aan/aan_bk.htm

Acta Anatomica

Acta Anat 1998;163:125–126

Preface

Are motor patterns evolutionarily conservative? Does the emergence of new mechanisms, such as in feeding, respiration or locomotion, depend primarily upon changes of motor patterns, upon changes of mechanisms (bone-muscle constructions), or upon a combination of the two? Is it possible to trace evolutionary changes in the organization of the neural substrate of premotor and motor systems? These and several other questions were the subject of a symposium on Brainstem (Pre)motor Mechanisms and the Stereotypy of Motor Patterns held during the 5th International Congress of Vertebrate Morphology (Bristol, July 1997). Five speakers discussed various systems and different approaches. Thus Grillner [1997] discussed the evolution of the neural substrate in control of locomotion. The point to make was that, despite the varied types of motor patterns in the various groups of vertebrates, the underlying neural control system has a basically similar design: from diencephalic and mesencephalic locomotor centers through reticulospinal connections to pattern-generating networks. Even though different groups of muscles may be involved, the patterns of activation of these muscles have striking similarities. Feldman [1997] elaborated on the organization and histochemical characteristics of the postulated respiratory center in the ventrolateral brainstem, the pre-Bötzinger complex. The properties of interconnected pacemaker cells in this complex appear critical for the generation of rhythm and pattern [also Rekling and Feldman, 1998].

Several of these topics return in the three contributions published in this special issue of *Acta Anatomica.* A common element of these papers is that emphasis is on various aspects of feeding and a few closely related mechanisms, but mainly mechanisms in tetrapods are discussed. This does not imply that fishes are less interesting in this respect. For instance, a great variety of feeding mechanisms can be distinguished among fishes and there has been quite a lot of discussion about the question of whether motor patterns are evolutionarily conservative or not: do new mechanisms evolve by changing the peripheral mechanisms (bone-muscle constructions) or do changes occur, primarily due to changing motor patterns [e.g., Lauder, 1990; Galis et al., 1994]. It can be questioned whether this is a matter of a real contradiction or rather a difference in emphasis. A more profound analysis including both peripheral mechanisms and central organization of (pre-)motor centers is needed to provide an answer.

Such an approach is presented in the study by Dicke et al. [1998]. These authors compare the neural systems in control of tongue mechanisms between representatives of the Anura and the Urodeles. The mechanisms of tongue ejection differ between the two groups. What are the differences in the organization of the neural systems? Do central pattern generator circuits for feeding evolve conservatively and is flexibility dependent on the influence of higher centers and/or sensory input [e.g. discussion in Nishikawa et al., 1992]? Dicke et al. [1998] point out that the neural substrate possesses largely a common pattern of organization at the higher levels of organization, the main difference being a difference of the functional characteristics of the neural substrate at the 'executive' level of the pattern generation.

The papers of Bout [1998] and Dubbeldam [1998] discuss the flexibility of jaw and tongue mechanisms and the organization of the reticular premotor systems in control of these mechanisms, respectively. The focus of the first paper

KARGER

Fax+41 61 306 12 34
E-Mail karger@karger.ch
www.karger.com

Accessible online at:
http://BioMedNet.com/karger

is on mechanisms in reptiles and birds, with some comments on the situation in fishes and mammals. The starting point is the question of whether a common basic movement pattern exists. What kinds of modulations can be expected, what are the consequences for the concept of the central pattern generator? Quite a lot of work has been done on jaw and tongue mechanisms in reptiles [see Bout, 1998], but little is known about the organization of reticular premotor systems for oral functions in these animals [for a recent overview on the organization and connections of the reptilian reticular formation, see ten Donkelaar, 1998]. Therefore, emphasis in the discussion on premotor systems [Dubbeldam, 1998] is on the avian situation and the comparison with mammals. The aim of that paper is to provide an overview of the organization of various, partly overlapping premotor systems in the medullary reticular formation, including those for locomotion and vocalization plus respiration.

The picture evolving from these presentations is that of a stable organization of the neural substrate which, however, does allow a high degree of flexibility of motor patterns enabling animals to respond adequately to an ever-changing environment. These papers may present some answers to the questions asked above, but are also meant as contributions to the ongoing discussions about the organization of motor and premotor systems. The collision of current ideas may help to generate new and more encompassing concepts and hypotheses.

J.L. Dubbeldam, Leiden

References

Bout, R.G. (1998) Complex movement patterns: Modifiability and constraints. Acta Anat *163:* 144–156.

Dicke, U., G. Roth, T. Matsushima (1998) Neural substrate for motor control of feeding in amphibians. Acta Anat *163:* 127–143.

Dubbeldam, J.L. (1998) The neural substrate for ‘learned’ and ‘non-learned’ activities in birds. A discussion of the organization of bulbar reticular premotor systems with side-lights on the mammalian situation. Acta Anat *163:* 157–172.

Feldman, J.L. (1997) Neural control system for breathing in vertebrates: From fish to cat. J Morphol *232:* 254.

Galis, F., A. Terlouw, J.W.M. Osse (1994) The relation between morphology and behaviour during ontogenetic and evolutionary changes. J Fish Biol *45(suppl A):* 13–26.

Grillner, S. (1997) Neural control system for locomotion in vertebrates: From cyclostomes to man. J Morphol *232:* 260.

Lauder, G.V. (1990) Biomechanics and evolution: Integrating physical and historical biology in the study of complex systems; in Rayner, J.V.M., R. J. Wootton (eds): Biomechanics in Evolution. SEB Seminar Series 36. London, Cambridge University Press, pp 1–20.

Nishikawa, K.C., C.W. Anderson, S.M. Deban, J.C. O’Reilly (1992) The evolution of neural circuits controlling feeding behavior in frogs. Brain Behav Evol *40:* 125–140.

Rekling, J.C., J.L. Feldman (1998) Pre-Bötzinger complex and pacemaker neurons: Hypothesized site and kernel for respiratory rhythm generation. Annu Rev Physiol *60:* 385–405.

ten Donkelaar, H.J. (1998) Reptiles; in Nieuwenhuys, R., H.J. ten Donkelaar, C. Nicholson (eds): The Central Nervous System of Vertebrates. Berlin, Springer, vol 2, pp 1315–1524.

Acta Anatomica

Acta Anat 1998;163:127–143

Received: June 30, 1998
Accepted: November 10, 1998

U. Dicke
G. Roth
T. Matsushima

Brain Research Institute,
University of Bremen, Germany

Neural Substrate for Motor Control of Feeding in Amphibians

Key Words
Descending pathways
Premotor/motor centers
Optic tectum
Hypoglossal motor nucleus
Medulla oblongata

Abstract
Descending pathways to premotor/motor centers and their cell groups of origin were studied by means of retrograde biocytin tracing experiments in the frog *Discoglossus pictus* and the plethodontid salamander *Plethodon jordani,* which differ remarkably in the structure and function of their feeding apparatus and their feeding strategy. Labeled neurons were found in 30 major cell groups located in the telencephalon, diencephalon, synencephalon, mesencephalon and rhombencephalon. The number and distribution of nuclei are very similar in both species. Furthermore, the descending pathways of these groups of neurons take the same courses inside the medulla oblongata. Axons of most nuclei descend either in the ventromedial or ventrolateral medulla oblongata, and it is concluded that the spatial arrangement of pathways is identical in the species studied. Bilateral electrical stimulation of the optic tectum of the plethodontid salamander *Hydromantes italicus* elicited strong discharges of short latencies in the hypoglossal nerve. In most hypoglossal motor neurons, excitatory postsynaptic potentials (EPSPs) of short latencies followed paired shocks applied at intervals as short as 3 ms, but showed temporal and spatial facilitation, suggesting that the EPSPs include mono- as well as polysynaptic components. In the ventral white matter, orthodromic single units were found that are candidates for excitatory reticular interneurons. These properties of tectal descending pathways in salamanders strongly differ from those found in toads. Differences in feeding behavior and its control by the premotor/motor networks between the species investigated do not appear to result from anatomically altered input or from a different organization of descending pathways to these premotor/motor centers, but rather from differences in local properties of reticular premotor networks as well as from different effects of neuromodulatory systems.

Abbreviations used in this paper:
1sp = First spinal nerve; 2sp = second spinal nerve; Amg = amygdala; Apo = area preoptica; Ca = commissura anterior; Ch = commissura hippocampi; Cho = chiasma opticum; cOT = contralateral optic tectum; Cp = commissura posterior; Cpo = Commisura postoptica; EPSP = excitatory postsynaptic potential; Hab = habenula; Hth = hypothalamus; I = olfactory nerve; II = optic nerve; III = oculomotor nerve; iOT = ipsilateral optic tectum; IPSP = inhibitory postsynaptic potential; IV = trochlear nerve; IX-X-XI = glossopharyngeal, vagal, accessory nerve; Mo = medulla oblongata; Ncb = nucleus cerebelli; Ndh = nucleus dorsalis hypothalami; Nflm = nucleus fasciculi longitudinalis medialis; Nlpd = nucleus lateroposterodorsalis thalami; Nlpv = nucleus lateroposteroventralis thalami; Nmg = nucleus magnocellularis area preoptica; Nptp = nucleus pretectalis profundus; Npts = nucleus pretectalis superficialis; Nr = nucleus ruber; Nri = nucleus reticularis isthmi; Nrl = nucleus reticularis lateralis; Nrm = nucleus reticularis medius; Nrs = nucleus reticularis superior; Nsc = nucleus suprachiasmaticus; Ntgd = nucleus tegmentalis dorsalis; Ntgv = nucleus tegmentalis ventralis; Nvt = nucleus ventralis thalami; OT = optic tectum; PB = phosphate buffer; Pd = pallium dorsale; Pl = pallium laterale; Pm = pallium mediale; Sp = septum; St = striatum; Tg = tegmentum; To = tectum opticum; Tp = tuberculum posterius; Ts = torus semicircularis; V = trigeminal nerve; VII = facial nerve; VIII = vestibulocochlear nerve.

KARGER

Fax + 41 61 306 12 34
E-Mail karger@karger.ch
www.karger.com

0001–5180/98/1633–0127$17.50/0

Accessible online at:
http://BioMedNet.com/karger

Dr. Ursula Dicke
Brain Research Institute, University of Bremen
D–28334 Bremen (Germany)
Tel. +49 421 2183140, Fax +49 421 2184549
E-Mail dicke@zfn.uni-bremen.de

Introduction

Frogs and salamanders differ considerably with respect to the anatomy and functional morphology of their feeding apparatus and their visually guided feeding strategies, the set of motor nuclei in the medulla oblongata and rostral spinal cord involved in feeding behavior as well as their cytoarchitecture, and finally the gross anatomy of visual centers involved in visual control of feeding [Roth et al., 1990]. While most frogs use tongue muscles for prey capture by rotating them over the rim of the lower jaw, in salamanders the entire hyobranchial apparatus is moved out of the mouth, with the tongue pad at its tip. In groups of salamanders of the family Plethodontidae, i.e., in the Bolitoglossini, this feeding apparatus is greatly elongated forming a projectile tongue that in its extreme has a reach equal to snout-vent length [Roth, 1987; Deban et al., 1997]. The differences between frogs and salamanders with respect to the feeding apparatus imply that in salamanders the nuclei of the IXth and Xth cranial nerves (glossopharyngeal and vagal nerve) are essential for feeding by controlling the tongue protractor muscle subarcualis rectus, while in frogs they are not involved in feeding at all, but in breathing and calling [Duellman and Trueb, 1986; Roth et al., 1988]. In frogs, the hypoglossal/1st spinal motor nucleus is involved both in protraction and retraction of tongue muscles, while in salamanders it serves only the tongue retractor muscle [Wake et al., 1988; Roth et al., 1990]. Finally, the optic tectum as the main center for visual guidance of behavior in the context of object recognition and localization is multilaminated in frogs, while it is essentially bilaminated in salamanders [Székely and Lázár, 1976; Roth, 1987]. The question arises whether these differences can – or cannot – be correlated with morphological and physiological differences inside the brains and rostral spinal cords of frogs and salamanders.

In order to answer this question, we studied the descending pathways and their cells and nuclei of origin to premotor/motor centers in a 'representative' frog, *Discoglossus pictus,* and a 'representative' salamander, *Plethodon jordani.* Similar studies have been carried out in our laboratory in the frog *Eleutherodactylus coqui* and the bolitoglossine salamander *Hydromantes italicus.* The findings resulting from these studies confirm those in *Discoglossus* and *Plethodon,* but are not presented here. Studies have been carried out in the frogs *Xenopus* and *Rana* and the salamander *Salamandra* using either degeneration technique or HRP or cobaltic lysine as tracers applied to the entire half of the medulla spinalis [Ten Donkelaar et al., 1981; Ten Donkelaar, 1982; Tóth et al., 1985; Naujoks-Manteuffel et al., 1988; Naujoks-Manteuffel and Manteuffel, 1988]. Descending pathways to the medulla oblongata have been investigated only in *Salamandra* by means of HRP application to the entire half of the medulla oblongata [Naujoks-Manteuffel and Manteuffel, 1988; Naujoks-Manteuffel et al., 1988]. In our study, we used the tracer biocytin, which is far more sensitive than the other tracers, and we used in vitro whole-brain preparations that allow a more precise application of the tracer.

Early models of visual guidance of feeding behavior in amphibians were based on the idea that specialized tectal 'prey recognition neurons' release the feeding response more or less directly [Ewert, 1974]. This idea was subsequently replaced by a parallel-processing concept according to which separate visual pathways carrying different kind of information about a prey object descend to premotor reticular centers, where they converge with each other as well as with nontectal descending pathways and then influence the motor centers related to feeding [Ewert, 1984, 1989; Satou and Ewert, 1985; Roth, 1987; Dicke and Roth, 1994]. Neurophysiological studies on reticular neurons in toads [Ewert et al., 1990; Schwippert et al., 1990] and salamanders [Heimbuch and Dicke, 1998] revealed highly complex response properties at visual and electrical stimulation. In order to understand these findings, it is of prime importance to have exact knowledge about the way in which descending pathways influence motor neurons related to feeding. Existing studies concern the interaction between the optic tectum and premotor and motor centers related to feeding; comparable studies on other descending pathways are still lacking. A study in the Japanese toad *Bufo japonicus* by one of the coauthors [Satou et al., 1985; Matsushima et al., 1986] had demonstrated that tectal neurons influence hypoglossal motor neurons mostly polysynaptically and that first responses of hypoglossal neurons to electrical stimulation of the optic tectum are always inhibitory.

Comparable data on the tectoreticular system in salamanders were still lacking. Accordingly, in the tongue-projecting salamander *H. italicus* we investigated physiological properties of the ipsi- and contralateral tectobulbospinal pathways by recording the types and latencies of responses of hypoglossal motor neurons of the hypoglossal nerve as well as units presynaptic to these neurons upon stimulation of the optic tectum. As will be shown in this article, these experiments revealed major differences among frogs/toads and salamanders in the way how the tectum influences premotor and motor centers related to feeding.

Material and Methods

Tracer Experiments

Descending projections to the medulla oblongata in the frog *D. pictus* (n=6) and the plethodontid salamander *P. jordani* (n=6) were investigated by means of in vitro retrograde tracing experiments. Animals were anesthetized in a 0.1% MS-222 solution (Sigma) and perfused with ice-cold oxygenated Ringer solution. The brain was immediately removed, and the tracer biocytin (Sigma) was applied to the ventromedial or ventrolateral medulla oblongata (for every site 3 animals of each species) at the level of the VIIth cranial nerve (fig. 4) after lesioning the site of application with the tip of a microelectrode. Brains were kept in oxygenated Ringer solution overnight at 4 °C and then fixated in phosphate buffer (PB: 0.1 m, pH 7.2) containing 1% paraformaldehyde and 2% glutaraldehyde. Brains were embedded in agar-agar (4.5% in PB) and cut on a vibratome into 40-μm-thick sections. Biocytin was localized by means of an avidin-biotin-horseradish peroxidase complex using DAB as chromogen (Vector Lab.). Identification of labeled cell groups was based on light-microscopic examination; the criteria for identification included the location of labeled neurons and their relationship to neighboring cell groups and fiber tracts, packing density and shape of somata as well as pattern of dendritic arborization. In each nucleus, the diameter of labeled somata was measured, and mean and standard deviation were calculated. The nomenclature of nuclei was adopted from Herrick [1948], Roth [1987] and Opdam and Nieuwenhuys [1976] for salamanders, and Neary and Northcutt [1983] for the diencephalon and Nieuwenhuys and Opdam [1976] for the brainstem of frogs.

Recording Experiments

The recording experiments were carried out in the plethodontid salamander *H. italicus* (n=10). Animals were anesthetized as described above, immobilized by intraperitoneal injection of succinyl choline (Asta; 0.1–0.2 ml, 0.5% in saline) and fixed on a head holder. Branches of the hypoglossal nerve were bilaterally exposed at the gular region, and exposed nerves were hooked on bipolar silver wire electrodes for electrical stimulation and nerve discharge recording. Optic tectum, medulla oblongata and rostral spinal cord were dorsally exposed by removing the overlying cranium and vertebrae, respectively. Bipolar stainless steel electrodes (exposed tip length: 100 μm, tip distance: 200–300 μm) were inserted into each tectal hemisphere at a depth of 250–350 μm. Rectangular current pulses (amplitude: 0.3–10 V, duration: 0.1 ms for hypoglossal nerve; 150 μA or less, and 0.1–0.5 ms for optic tectum) were applied. Field potentials were recorded using glass micropipettes filled with 2 *M* NaCl (DC resistance 2–12 MΩ), and intracellular potentials were recorded using 2 *M* K-citrate glass micropipettes (DC resistance 40–60 MΩ). For recording nerve discharges, the hook electrodes were connected to a DC-differential amplifier. These potentials were further amplified by an oscilloscope preamplifier and stored in a computer at a sampling rate of 20 or 50 kHz through a 12-bit A/D converter.

Results

Tracer Experiments

In all animals studied, neurons were well stained from the telencephalon to the medulla oblongata after application of biocytin to the ventromedial and ventrolateral white matter of the medulla oblongata at the level of the VIIth nerve. In both species, a minimum of 30 major cell groups giving rise to descending pathways could be identified. Examples for the distribution of labeled somata in the different parts of the brain are given in figures 1 and 2. Labeled cell groups were not further subdivided, e.g., into a medial and a lateral nucleus within the ventral thalamus. Some of these groups are known to consist of subgroups, but these can be distinguished only by using other techniques such as intracellular injection of tracers or double-labeling experiments [Dicke, 1998; Luksch and Walkowiak, 1998; Roth et al., 1998].

The number and distribution of labeled nuclei constituting the descending pathways to the medulla oblongata were very similar in *D. pictus* and *P. jordani* (fig. 1, 2). Therefore, the data from both species concerning the distribution of labeled cell groups are presented together, if not mentioned otherwise.

In all animals studied, the site of lesioning extended about 200 μm in the fiber layer of the ventromedial or ventrolateral medulla oblongata. The tracer did not diffuse into unlesioned surrounding regions, e.g. it did not cross the midline. One example for the extent of the application site in *Plethodon* and *Discoglossus* is given in figure 5.

Somata of labeled neurons have a mean diameter of 9.3±1.4 μm in *D. pictus* and of 13.3±1.7 μm in *P. jordani.* In both species, somata with diameters above average belonged to the following cell groups or nuclei: striatum, magnocellular nucleus of the preoptic area, suprachiasmatic nucleus, superficial pretectal nucleus, nucleus of the medial longitudinal fasciculus, trigeminal mesencephalic nucleus, and torus semicircularis ipsilateral to the application site, and red nucleus, optic tectum and lateral reticular zone contralateral to the application site.

Telencephalon. Numerous neurons were labeled in the ipsilateral dorsal and ventral striatum after tracer application in the ventrolateral medulla oblongata (fig. 3A, B). With average diameters of 12.5 μm in the frog and 14.5 μm in the salamander, striatal neurons belong to the group of nuclei characterized by large somata. In one salamander specimen, labeled neurons were found in the transition zone between the striatum and the lateral pallium as well. In frogs and salamanders, a few labeled neurons were situated in the ipsilateral amygdala.

Diencephalon. Labeled neurons were found in the ipsilateral anterior preoptic area after medial and lateral tracer application. In addition, neurons of the magnocellular nucleus of the preoptic area were labeled ipsilaterally only after lateral tracer application (fig. 3C, D). These neurons likewise have large somata (10.5 μm in the frog and 15.5 μm in the

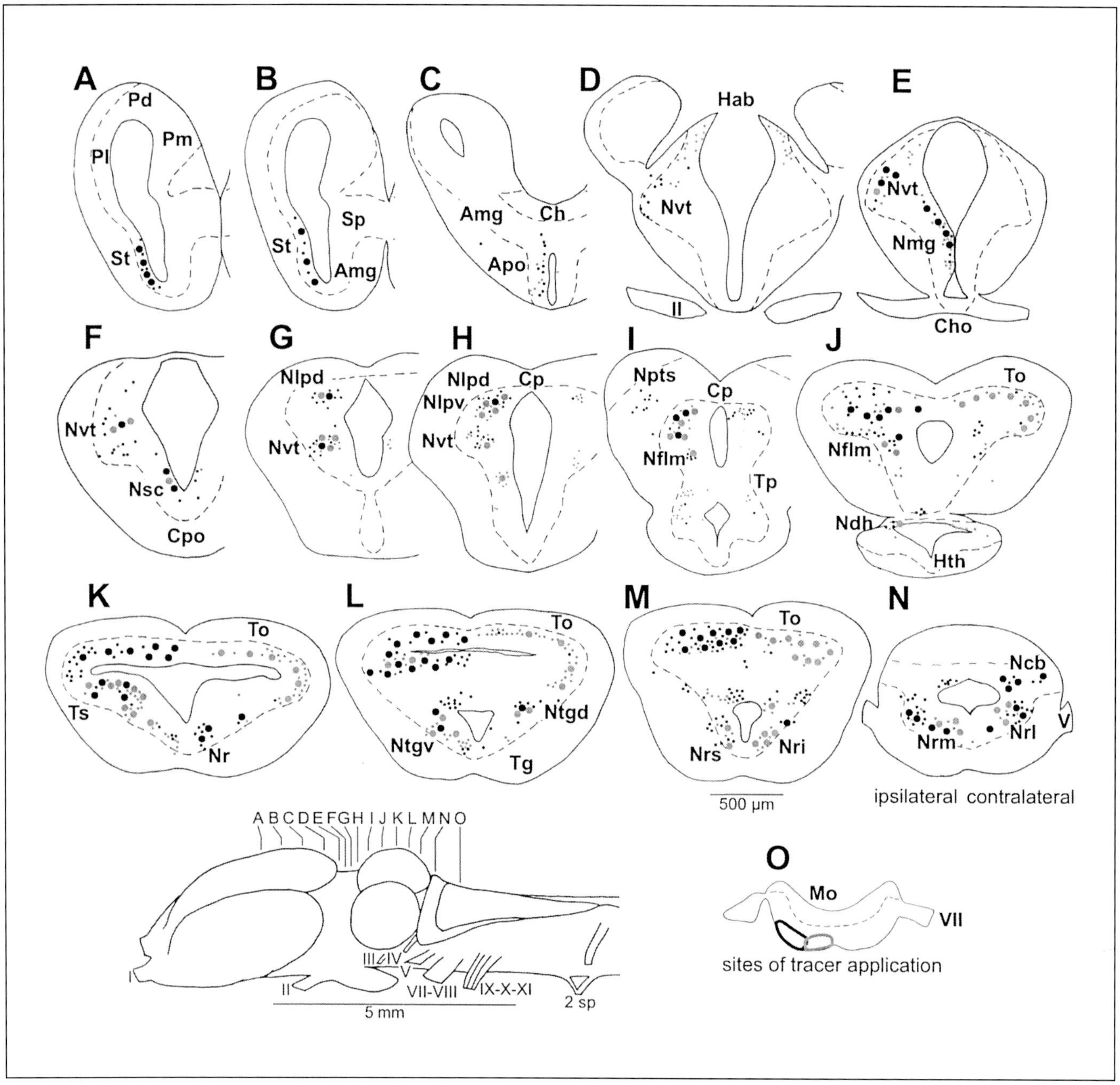

Fig. 1. Representative transverse sections of the brain of *D. pictus* (**A–O**) illustrating the distribution of labeled neurons. The results from two cases of application of biocytin, one to the ventromedial and the other to the ventrolateral site of the medulla oblongata, are plotted into the same set of sections. Small dots represent one single labeled soma, large dots ten labeled somata. For abbreviations see list of abbreviations on page 127.

salamander). Again ipsilaterally, neurons were labeled in the ventrolateral and ventromedial thalamic nucleus forming a more or less continuous rostrocaudal band. Neurons situated laterally were stained predominantly after ventrolateral tracer application, while ventromedial application resulted in labeling of neurons both medially and laterally (fig. 3C–F). After ventromedial tracer application, in the frog a number of labeled neurons were found in the rostral portion of the ipsilateral dorsal thalamus, while in the salamander very few were found here. In the frog, only few

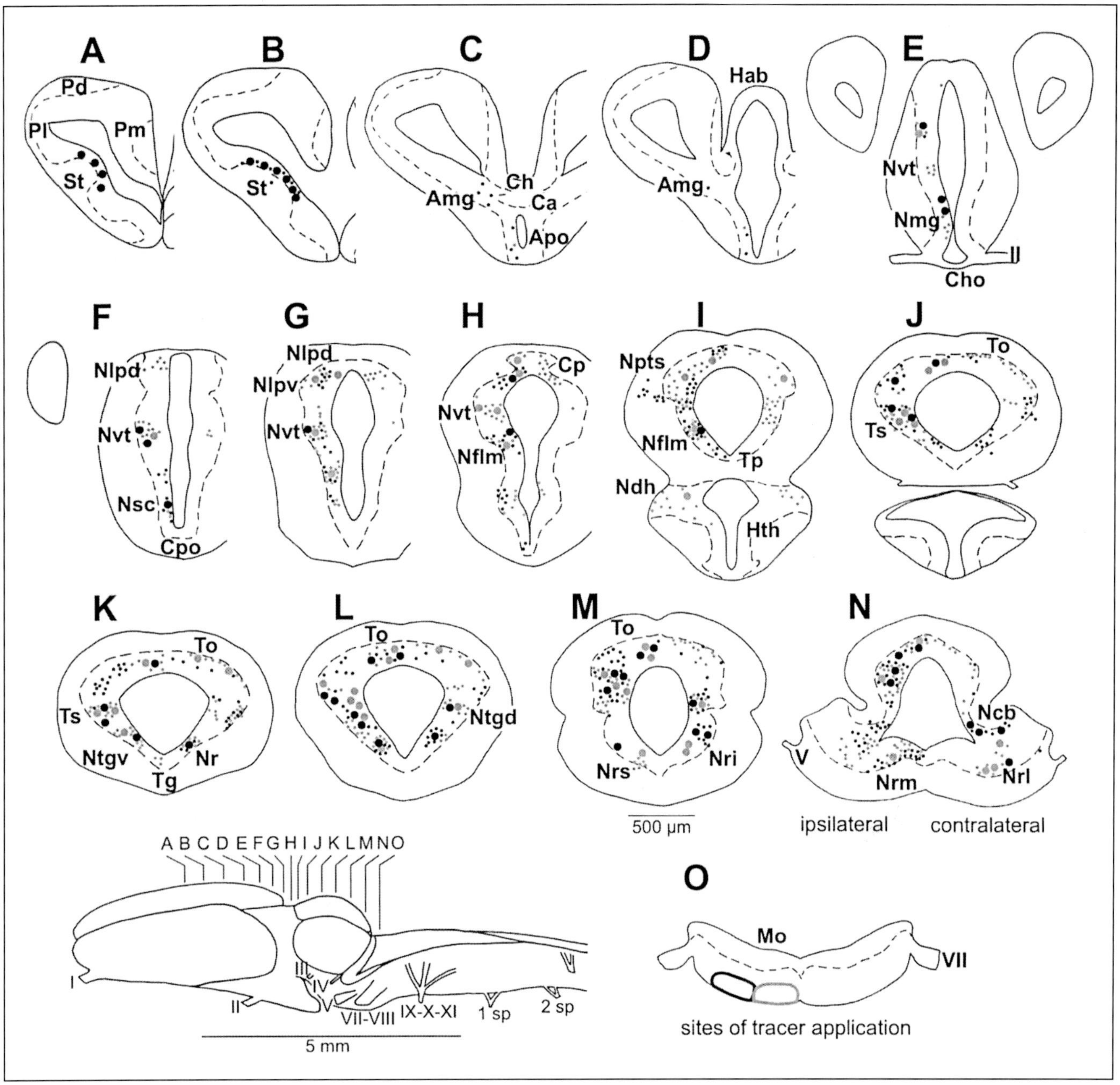

Fig. 2. Representative transverse sections of the brain of *P. jordani* (**A–O**) illustrating the distribution of labeled neurons. Results from two cases of tracer application, one to the ventromedial and the other one to the ventrolateral medulla oblongata, are plotted into the same set of sections. Small dots indicate one single labeled soma, large dots ten labeled somata. For abbrevations see list of abbreviations on page 127.

neurons were found in the ipsilateral lateral thalamic nuclei, while many neurons were located in the caudally continuing lateral posterodorsal and posteroventral thalamic nuclei. In the central thalamic nucleus, only few labeled neurons were found. Neurons in the suprachiasmatic nucleus were labeled ipsilaterally after medial and lateral tracer application (fig. 3E, F); their large somata have mean diameters of 10.5 µm in frogs and 16.5 µm in salamanders. Ipsilaterally, neurons of the nucleus fasciculi longitudinalis medialis (Nflm) were labeled predominantly after medial tracer

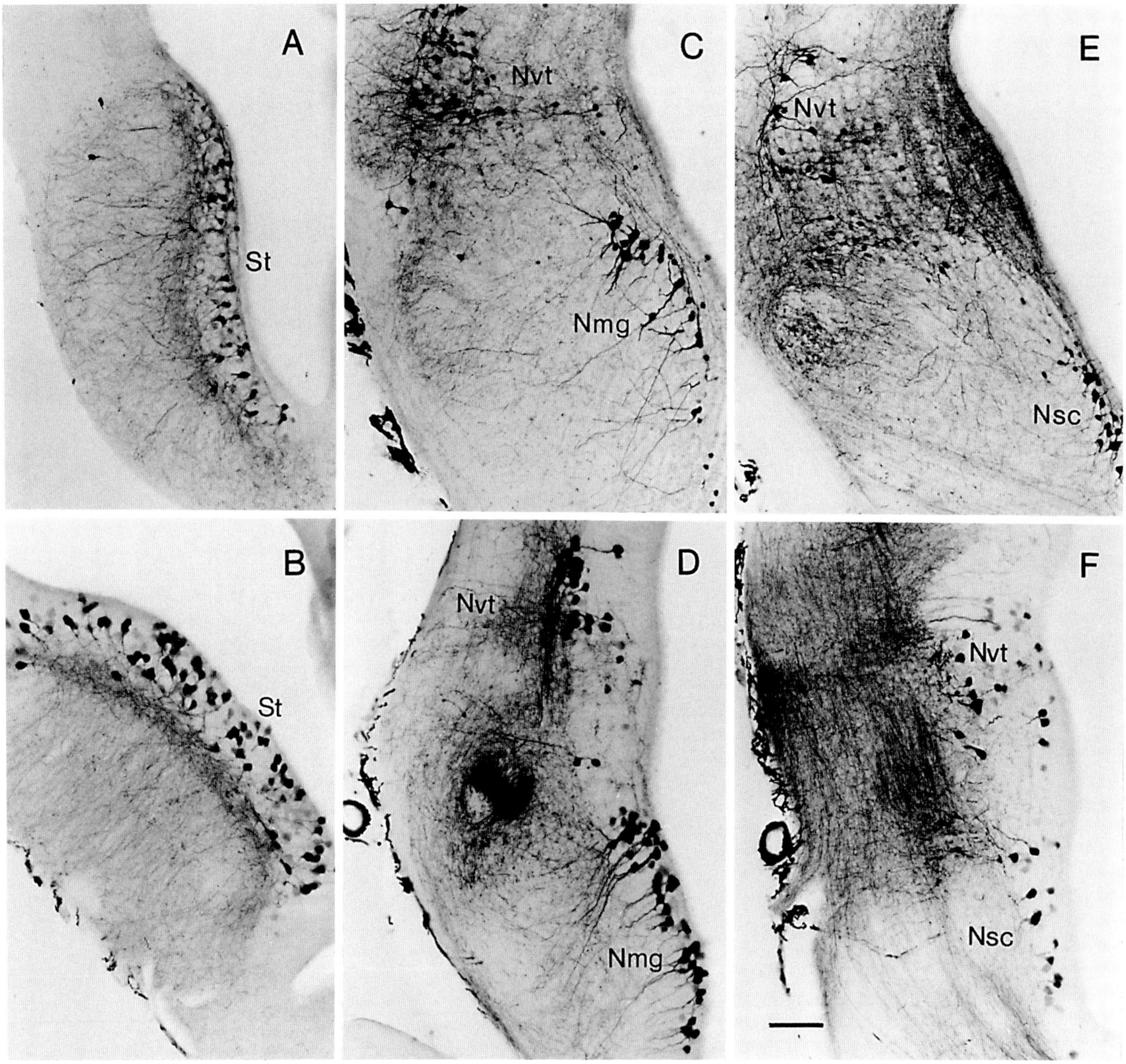

Fig. 3. Microphotographs of transverse sections of *D. pictus* (**A**, **C**, **E**) and *P. jordani* (**B**, **D**, **F**) showing neurons labeled ipsilaterally to the site of tracer application. In the telencephalon, neurons were labeled in the striatum (St) (**A**, **B**), and in the diencephalon in the ventral thalamus (Nvt) and the magnocellular nucleus of the preoptic area (Nmg) (**C**, **D**). They were stained after application of biocytin to the ventrolateral medulla. **E**, **F** Neurons in the ventral thalamus (Nvt) and in the suprachiasmaticus nucleus (Nsc) shown at the level of the postoptic commissure were labeled after ventromedial application. From this level of section to the caudal thalamus, neurons of the lateral thalamic nuclei are present. **F** Bar = 100 μm; it also applies to **A**–**E**.

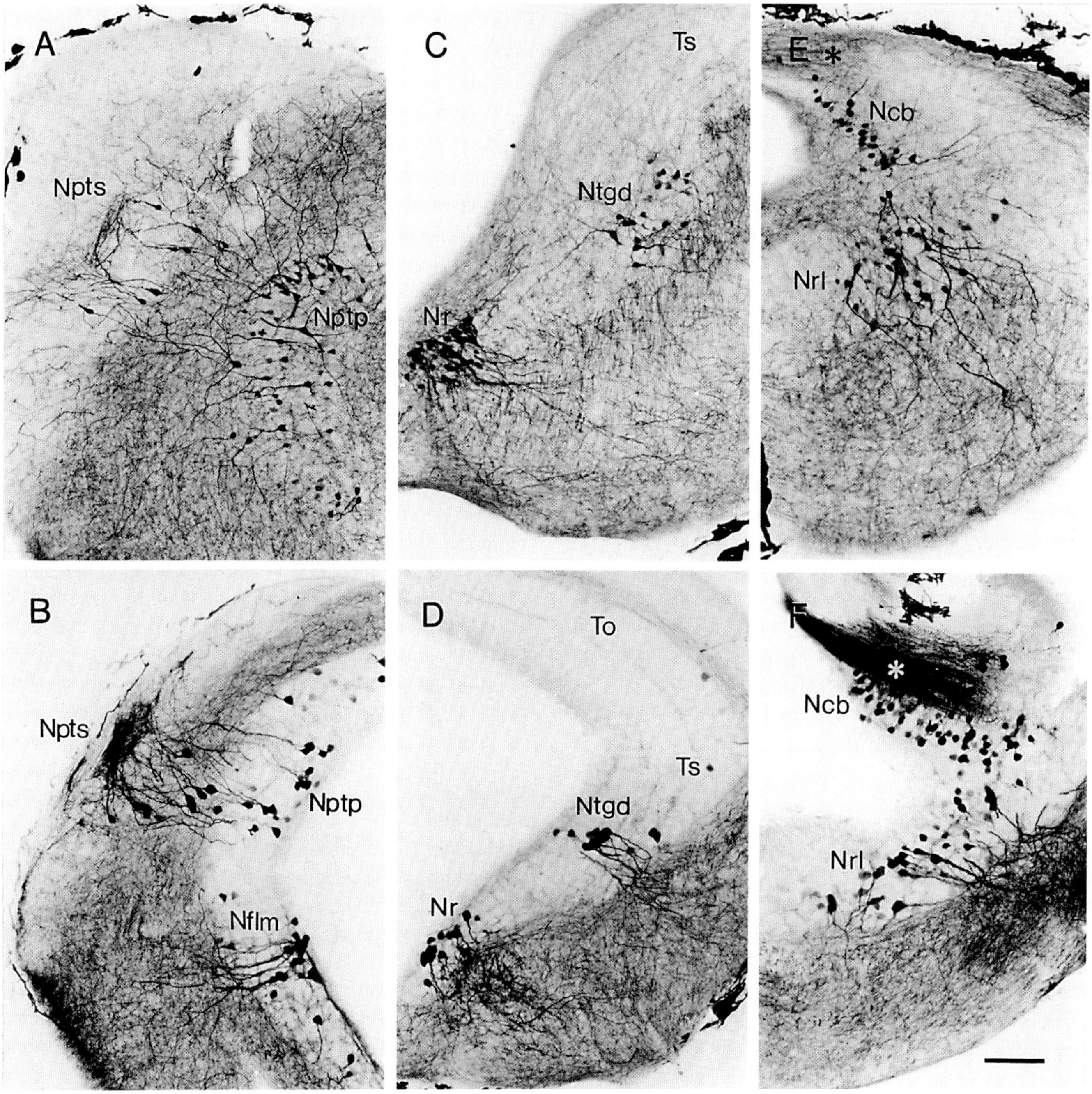

Fig. 4. Microphotographs of transverse sections of *D. pictus* (**A**, **C**, **E**) and *P. jordani* (**B**, **D**, **F**) showing labeled nuclei after tracer application to the ventrolateral medulla. **A**, **B** Neurons of the superficial and deep pretectal nucleus (Npts, Nptp) are labeled ipsilaterally to the site of application. Labeled neurons of the medial longitudinal fascicle (Nflm) are found in the transition zone between ventral thalamus and rostral tegmentum. **C**, **D** In the midbrain tegmentum, neurons of the anterior portion of the dorsal tegmental nucleus (Ntgd) and the posterior portion of the red nucleus (Nr) are found. They were labeled contralaterally to the site of application. **E**, **F** Labeled neurons of the cerebellum (Ncb) and of the lateral reticular formation (Nrl) are also found contralaterally to the site of application. Asterisks indicate the cerebellar commissure. To = Tectum opticum; Ts = torus semicircularis. **F** Bar = 100 μm.

application; in two salamander specimens, a substantial number of somata were likewise labeled after lateral tracer application (fig. 4B). In the transition zone between caudal thalamus and rostral mesencephalon, numerous neurons were stained after medial tracer application. Neurons of the Nflm were easy to distinguish because of their large soma size (11 μm in the frog and 16 μm in the salamander). A number of neurons with smaller somata were located rostrodorsal as well as caudoventral to the Nflm. They probably belong to the nucleus of Darkschewitsch and the nucleus of Edinger-Westphal, respectively. Some labeled neurons were also found in the posterior tubercle. Cells were labeled in the pars dorsalis of the mid and caudal hypothalamus. They were found mainly ipsilaterally.

Synencephalon. Rostral to and at the level of the posterior commissure (Cp), labeled neurons were found in the dorsal gray matter. These neurons either belong to the caudal part of the lateral posterodorsal thalamic nucleus or constitute the juxtacommissural nucleus described by Puelles et al. [1996]. They were labeled predominantly on the ipsilateral side. At the level of the Cp and inside the gray matter, neurons of the nucleus pretectalis profundus were labeled bilaterally after ventromedial and ventrolateral tracer application, with fewer neurons labeled contralaterally. Large spindle-shaped neurons (long diameters of 18 μm in the frog and 21 μm in the salamander) of the nucleus pretectalis superficialis were labeled only ipsilaterally and only after lateral tracer application (fig. 4A, B).

Mesencephalon. After ventromedial tracer application to the medulla oblongata, neurons were stained in the contralateral optic tectum (OT) in the frog and the salamander, and in the ipsilateral OT in addition in the salamander (fig. 5F, G). After ventrolateral application, numerous neurons were found in the ipsilateral OT (fig. 5A, B). Different types of tectal neurons were labeled contralaterally and ipsilaterally; somata were always larger contralaterally (13 μm in the frog and 15.5 μm in the salamander) than ipsilaterally (10 μm in the frog and 14 μm in the salamander). Neurons of the mesencephalic nucleus of the trigeminal nerve were labeled only after ventrolateral application and dispersed in the ipsilateral optic tectum. Their somata are the largest of all neurons projecting to the medulla oblongata (22.5 μm in the frog and 25 μm in the salamander). After medial and lateral tracer application, many neurons were stained in the ipsilateral torus semicircularis (diameters: 13 μm in the frog and 16 μm in the salamander; fig. 5D, E). After ventromedial and ventrolateral tracer application, labeled neurons are found ipsilaterally throughout the entire dorsal tegmental nucleus, while contralaterally neurons were found more caudally (fig. 4C, D). Neurons in the ipsilateral ventral tegmental nucleus were stained after ventromedial tracer application (fig. 5D, E). After ventrolateral tracer application many neurons were labeled in the contralateral and a few in the ipslateral nucleus ruber (fig. 4C, D).

Rhombencephalon. The rostralmost groups of labeled neurons were found in the transition zone between tegmentum and rostral medulla oblongata. The isthmic reticular nucleus was labeled contralateral to the site of application. In the ipsilateral caudal tegmentum, neurons of the ventral tegmental nucleus and the superior reticular nucleus (Nrs) were labeled in close neighborhood, while more caudally the two nuclei are well separated. Neurons of the Nrs were located at the same position ipsilaterally as neurons of the reticular isthmic nucleus on the contralateral side and could be distinguished only because dendrites of the latter extend laterally and those of the Nrs neurons ventrally and ventrolaterally. The Nrs was continuous with the medial reticular nucleus. Neurons in the cerebellar nucleus were stained contralaterally only after lateral tracer application (fig. 4E, F). This was also the case for neurons labeled in the lateral reticular zone of the rostral medulla oblongata (fig. 4E, F). This group of neurons extend caudally at least to the level of the site of application.

Contralaterally, from the level of the entrance of the Vth cranial nerve to the caudal medulla, labeled neurons were also found in the dorsolateral edge of the area octavolateralis.

Ipsilaterally, at the level between the Vth and VIIth cranial nerves and close to the site of tracer application, labeled neurons were too numerous to be grouped into separate nuclei. The distribution of nuclei giving rise to the major pathways descending to the ventromedial or ventrolateral medulla oblongata in *Plethodon* and *Discoglossus* is summarized in figure 6.

Recording Experiments

Electrical stimulation of the ipsilateral or contralateral hemisphere of the optic tectum of *H. italicus* elicited strong

Fig. 5. Microphotographs of transverse sections of the midbrain of *D. pictus* (**A**, **D**, **F**, **H**) and *P. jordani* (**B**, **C**, **E**, **G**). **A**, **B** Neurons are labeled exclusively in the ipsilateral tectum opticum (To), and in the torus semicircularis (Ts) and tegmentum (Tg) after tracer application to the ventrolateral medulla (**C**). After ventromedial application (**H**), a large number of neurons in the torus and tegmentum were found ipsilaterally to the site of application (**D**, **E**), whereas contralaterally tectal neurons and a smaller number of neurons in the torus were labeled (**G**, **H**). Asterisks indicate labeled neurons in the ventral nucleus of the tegmentum (Tg). Arrowheads point to the midline, and arrows to the site of application. Bar = 200 μm (**A**) and 100 μm (**B–H**).

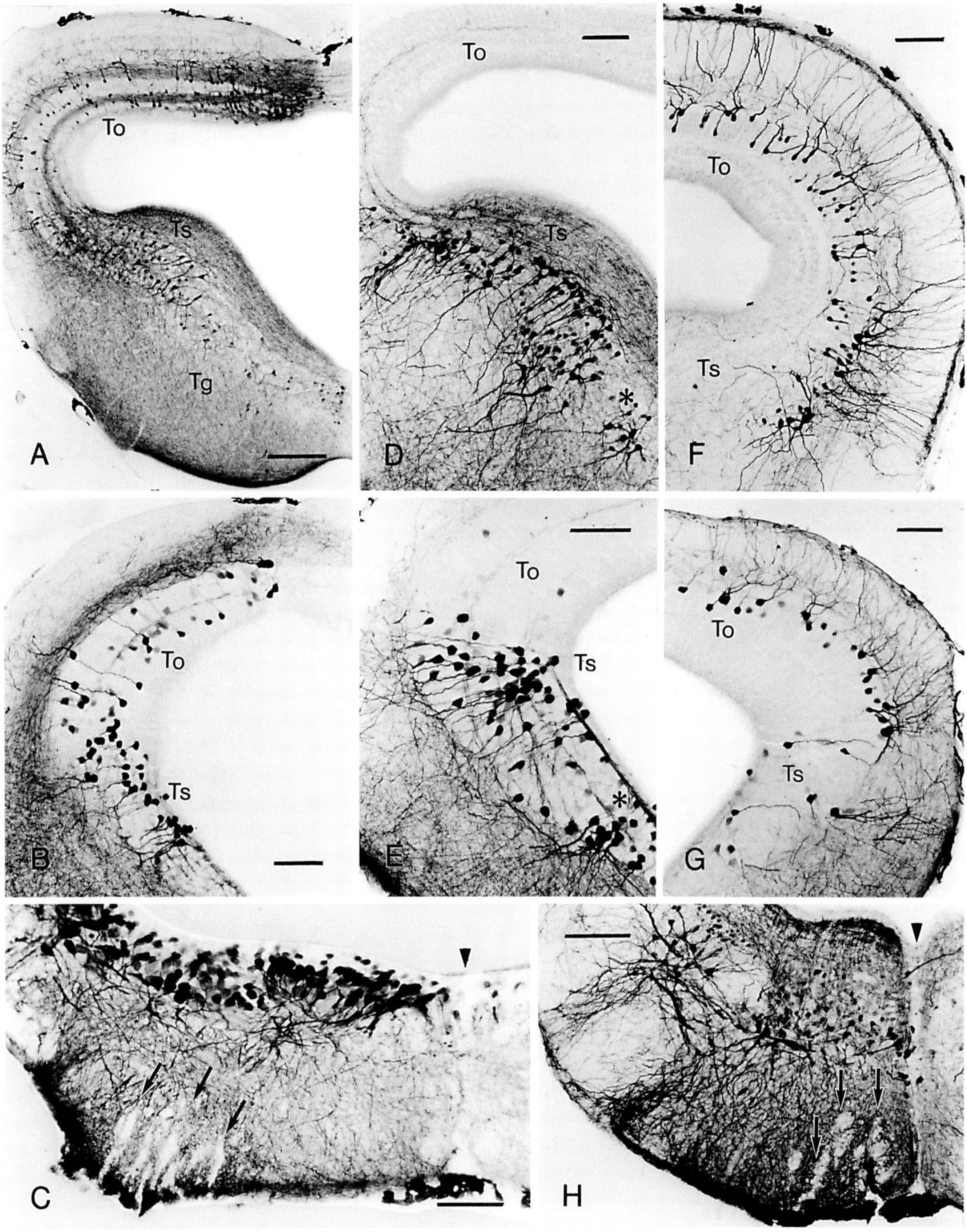
To
Ts
Tg
A
To
Ts
*
D
To
Ts
F
To
Ts
B
To
Ts
*
E
To
Ts
G
C
H
5

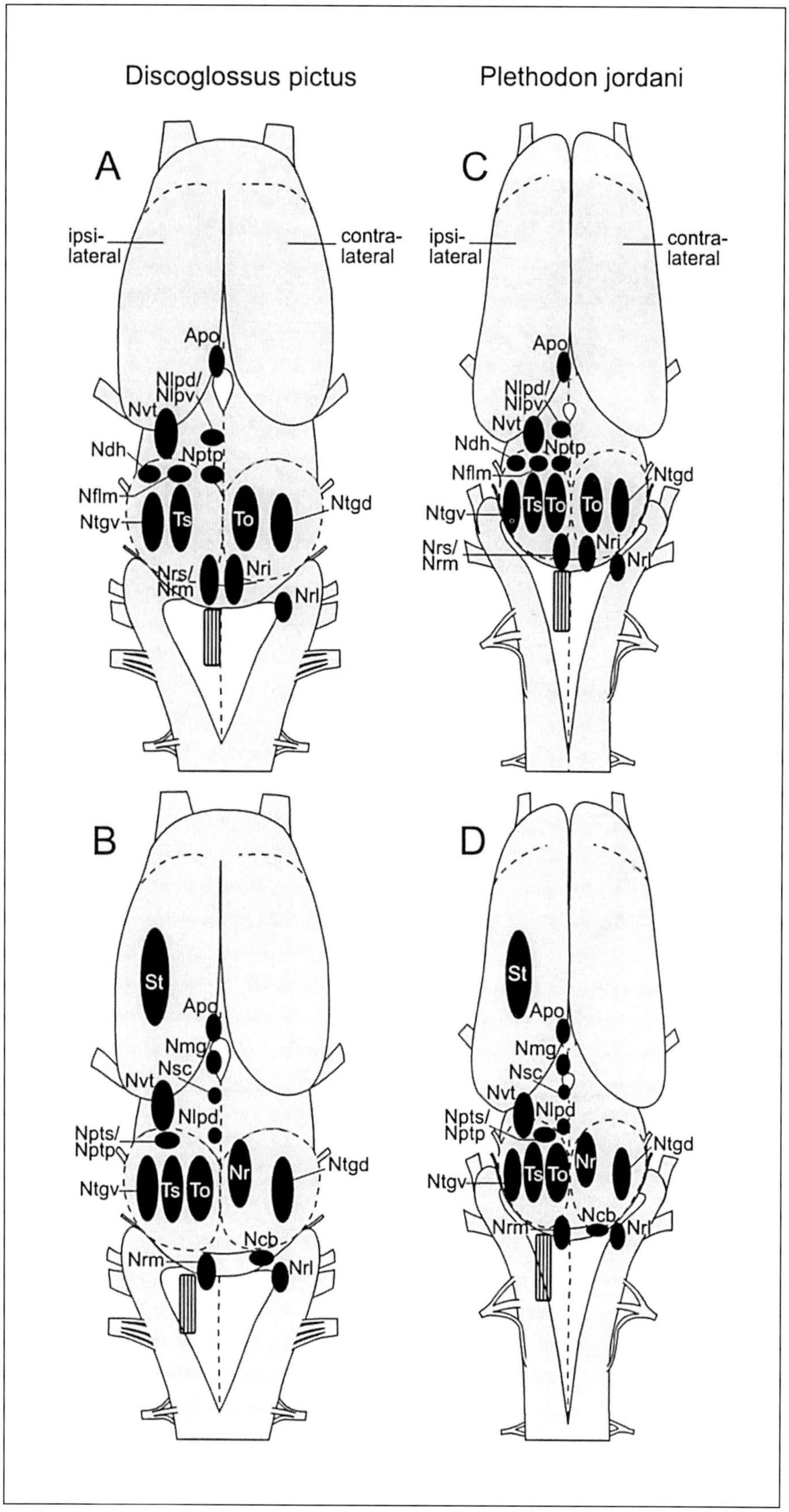

Fig. 6. Diagrams showing the distribution of the nuclei giving rise to the major descending pathways in *Discoglossus* (**A**, **B**) and *Plethodon* (**C**, **D**). Dorsal view of the brain. Black dots or areas represent the site of labeled nuclei; striped areas indicate the site of the descending tracts in the medulla oblongata. In **A** and **C**, nuclei with axons descending ipsi- or contralaterally to the ventromedial medulla oblongata are shown, while in **B** and **D** those with axons descending to the ventrolateral medulla are presented. For abbreviations see list of abbreviations on page 127.

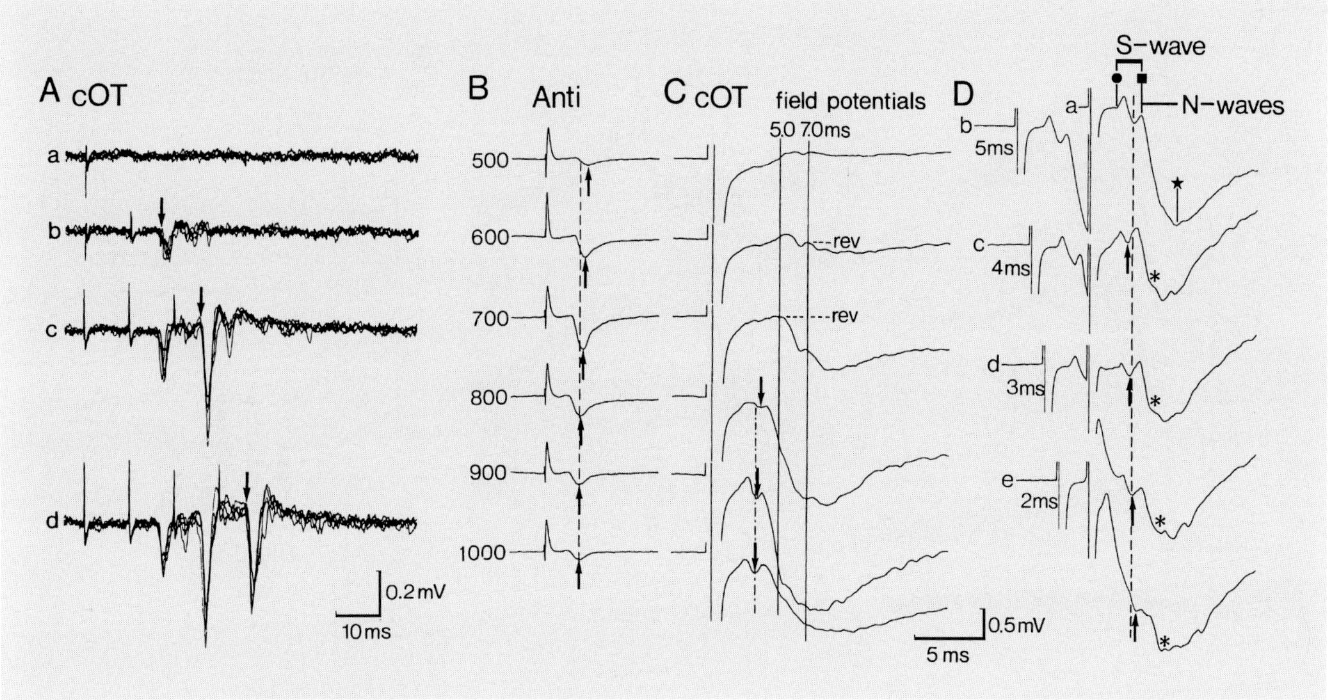

Fig. 7. A Responses of the hypoglossal nerve to electrical stimulation applied to the cOT (amplitude: 150 μA, duration: 0.5 ms). From line a–d, one to four successive shocks were applied at 10-ms intervals. Five superimposed traces are shown. Arrows indicate onset of responses caused by the last stimulus pulse. **B**, **C** Field potentials recorded near the hypoglossal nucleus. Anti- and cOT denote the responses to antidromic (supramaximal intensity) and cOT stimulation (150 A, 0.5 ms), respectively. Numbers left of the traces indicate the depth (μm) of recording sites. Averaged traces (n = 14) are shown. Arrows indicate peaks of the antidromic field potentials (left) and of the S-wave (right). The reversing points of the cOT fields (at 5 and 7 ms from stimulation onset) are indicated by rev. **D** Field potentials elicited by paired cOT shocks applied at short intervals (5–2 ms). Averaged traces (n = 14) are shown. Arrows indicate the S-waves and asterisks the N-waves elicited by the second shocks. Onset of S-waves (circle), onset (rectangle) and negative peak (star) of the N-waves are shown.

discharges of short latencies in the hypoglossal nerves on both sides. There were no consistent differences between the responses to stimulation of the ipsilateral and contralateral optic tectum (iOT and cOT). Figure 7A (lines a–d) shows responses of hypoglossal nerves to cOT stimulation. Single shocks elicited no responses (fig. 7A, line a), while 2–4 successive shocks applied at intervals of 10 ms induced a series of fairly synchronized action potentials, which followed each stimulus pulse with short and constant delays (arrows in fig. 7A, lines b–d). These delays were always less than 10 ms; minimal delay times were 5.2 and 5.6 ms for iOT and cOT stimulation, respectively.

In *Hydromantes,* the hypoglossal motor nucleus extends from the 3rd root of the IXth to Xth complex in the medulla oblongata to the middle between the roots of the 1st and 2nd spinal nerves in the rostral spinal cord. In order to identify the intramedullary and intraspinal representatives of tectal descending volleys and to precisely localize them relative to this motor nucleus, we analyzed the field potentials. Electrodes were moved vertically through the cord 200–300 μm lateral to the midline in the caudal medulla oblongata. Extracellular potentials were recorded at steps of 25–10 μm in response to stimulation of the hypoglossal nerve as well as of the contralateral and ipsilateral optic tectum. Antidromic responses were composed of single spike-like negative waves (fig. 7B) and had maximal amplitudes at depths of 590–850 μm, corresponding to the site of the hypoglossal motor nucleus. The field potentials after stimulation of the contralateral optic tectum were composed of an initial spike-like wave (S-wave) followed by slow negative waves (N-waves). Amplitudes of the S-wave were largest at sites 100–250 μm, and those of the N-wave at 100–250 μm ventral to the hypoglossal nucleus. The S-wave followed paired shocks applied at very short in-

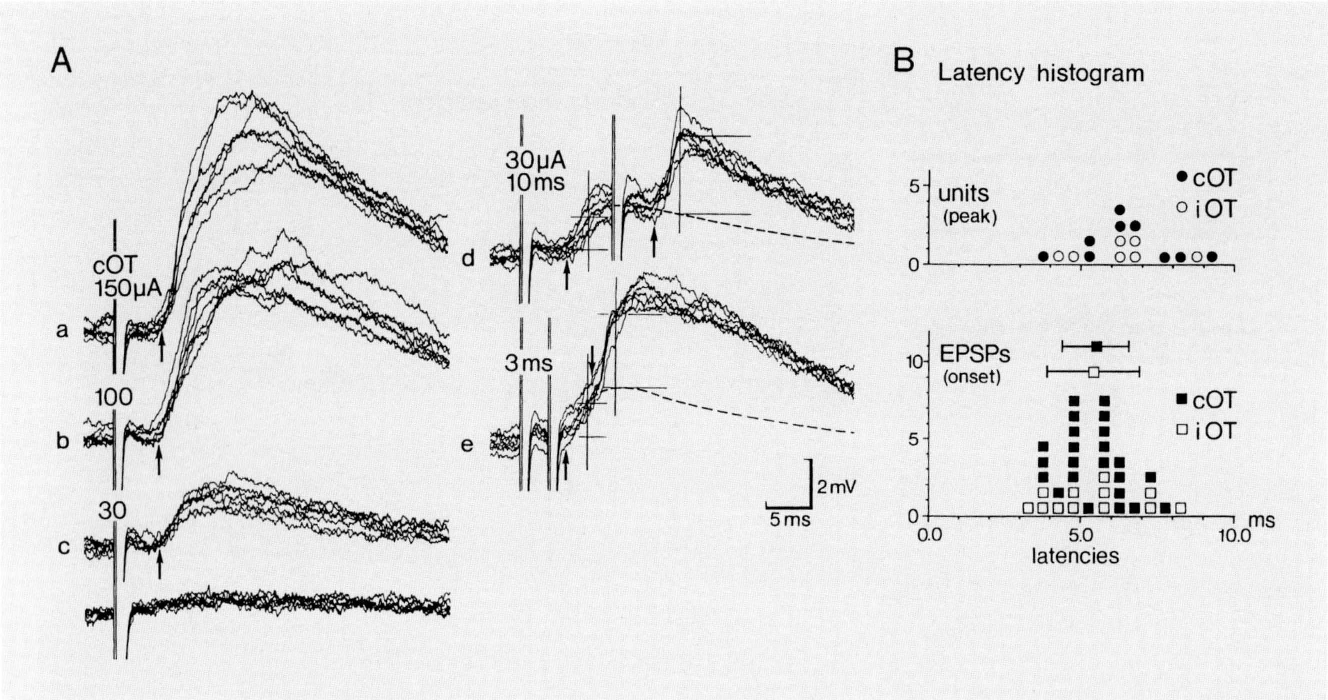

Fig. 8. **A** EPSPs in a hypoglossal motor neuron elicited by cOT shocks. Changes in stimulus amplitude (150–30 μA, lines a–c) did not result in detectable changes in the onset latencies of EPSPs (arrows). Seven superimposed traces are shown. **A** Traces below line c show field potentials recorded just outside the impaled cell (amplitude: 150 μA). The fastest components followed paired cOT shocks applied at short intervals (10 ms in line d, and 3 ms in line e). Dashed curves represent averaged responses to single cOT shocks (n = 14). Arrows indicate onset of EPSP responses; the second responses had slightly shorter latencies. Thin horizontal and vertical lines indicate the amplitude of EPSPs at 7.5 ms after each shock. The amplitudes of EPSPs to the second shocks (3.1 mV in line d, and 3 mV in line e; averaged value, n = 14) were larger than those of the first shocks (1.3 mV, n = 14). **B** Latency histogram measured from the onset of stimulation. Units (peak) show the latency histograms of extracellular single unit responses encountered in the white matter ventral to the hypoglossal nucleus; latencies of the positive peaks were measured. EPSPs (onset) show onset latencies of EPSPs in the hypoglossal motor neurons. Symbols and bars above the histogram indicate mean and standard deviation.

tervals (minimum: 2 ms) and almost disappeared at sites dorsal to the nucleus of the hypoglossal nerve (arrows in fig. 7C, D). On the other hand, the polarity of the N-waves was reversed at sites between 195 μm dorsal and 50 μm ventral to the hypoglossal nucleus. Shocks to the ipsilateral optic tectum induced slow negative waves without initial spike-like components, but the polarity of the slow waves usually was not reversed. These results suggest that the S-wave represents descending volleys of tectal efferents, and that the N-waves represent postsynaptic responses, presumably excitatory postsynaptic currents, induced by tectal volleys.

Our intracellular recordings from 28 neurons of the hypoglossal nerve revealed excitatory postsynaptic potentials (EPSPs) in response to tectal stimulation in most neurons. Single shocks applied to the ipsilateral optic tectum elicited EPSPs in 18 motor neurons, inhibitory postsynaptic potentials (IPSPs) in 2 motor neurons, EPSPs followed by IPSPs in 1 motor neuron, and no responses in 7 motor neurons. Under the same conditions, single shocks applied to the contralateral optic tectum elicited EPSPs in 28 motor neurons, IPSPs in 3 motor neurons, EPSPs followed by IPSPs in 1 motor neuron, and no responses in 4 motor neurons out of a total of 36 motor neurons examined. In most cases, those motor neurons that responded with EPSPs, IPSPs, or EPSPs-IPSPs to shocks applied to the ipsilateral tectum responded with the same type of postsynaptic potentials to shocks to the contralateral tectum. Figure 8A (lines a–e) shows examples of EPSPs. An increase in stimulus intensity resulted in larger amplitudes and longer peak latencies, but not in shorter onset latencies (fig. 8A, lines a–c). Furthermore, these EPSPs showed temporal and spatial facilitation. When submaximal stimulations were repeated at short intervals of 10 ms or less, the amplitudes of initial components of EPSPs were temporally facilitated (fig. 8A, lines d, e). In some other motor neurons, EPSPs were further spatially facilitated, when shocks to the ipsilateral and contralateral optic tectum were combined (data not shown).

In the white matter ventral to the hypoglossal nucleus, extracellular single units were encountered whose responses to tectal stimulation closely resembled the initial components of the EPSPs recorded in motor neurons. Among 15 extracellular units recorded, 3 units responded to shocks to the ipsilateral and 3 units to shocks to the contralateral optic tectum, while 9 units responded both to shocks applied to the ipsilateral and contralateral optic tectum. The latencies of their spike responses overlapped with the onset latencies of EPSPs recorded in motor neurons (fig. 8B); the earliest unit responses were almost coincident with the earliest onset latencies of the EPSPs in motor neurons, and also with the onset of slow negative waves (N-waves) elicited by tectal stimulation. In most of these units, spike responses followed each of successive stimulus pulses at intervals of 10 ms with short delays, usually less than 10 ms, similar to the EPSPs in motor neurons. Temporal and spatial facilitation was also observed. In 6 out of 9 units responding to bilateral OT shocks, spike responses were facilitated when iOT and cOT shocks were combined; latencies of the earliest spikes were shorter than those elicited by either iOT or cOT shocks.

Furthermore, tetanic stimulation of the optic tectum, which enhanced the nerve discharges of the hypoglossal nerves, also induced (1) enhancements in the amplitudes of N waves elicited by stimulation of the ipsilateral and contralateral optic tectum, (2) spike responses of premotor candidates, (3) peak amplitude of EPSPs in the motor neurons of the hypoglossal nerve, and (4) antidromic field potentials elicited by submaximal stimulation of the hypoglossal nerve.

Discussion

Distribution of Cell Groups Giving Rise to Descending Pathways

Our study shows that in the frog *D. pictus* and in the salamander *P. jordani,* descending projections to the medulla oblongata arise from at least 30 major cell groups situated in the telencephalon, diencephalon, synencephalon, mesencephalon and rhombencephalon. In both species, the number and distribution of nuclei are more or less identical, as is true for the set of nuclei characterized by large somata. The majority of descending nuclei project to the *ipsilateral* medulla oblongata. Axons of the different descending nuclei take similar courses inside the medulla oblongata, i.e., they run either in the ventromedial or ventrolateral white matter of the medulla oblongata. Also, the arrangement of pathways along the mediolateral axis is the same in the frog and the salamander species studied. Tracer application to the ventromedial or ventrolateral white matter of the medulla oblongata was likewise carried out in the frog *E. coqui* and the salamander *H. italicus* and yielded a labeling of the same groups of nuclei and pathways descending to the medulla oblongata [Dicke, 1997].

The results presented here demonstrate that the number of descending nuclei is larger than previously described in amphibians. Earlier studies in frogs using degeneration technique, HRP or cobaltic lysine as a retrograde tracer focused on pathways descending to the spinal cord and, thus, revealed only a subset of the nuclei described here [Tóth et al., 1985; Ten Donkelaar et al., 1981; Ten Donkelaar, 1982]. Naujoks-Manteuffel and Manteuffel [1988] applied HRP both to the medulla oblongata and spinalis in *Salamandra,* but the number of labeled cell groups as well as the number of labeled neurons was lower than in the present study. Also, the rostrocaudal extension of nuclei is larger in the present study. These differences certainly are due to the better quality of biocytin as a retrograde tracer and to the fact that using in vitro whole brains the tracer can be applied much more locally. Since in the mentioned previous studies the tracers had been applied to the entire half of the medulla oblongata and not to lateral or medial portions of it, the spatial arrangement of the descending pathways as reported here could not be discovered.

Recent tracer studies in the visual and auditory system of frogs and salamanders based on intracellular labeling of neurons demonstrate that the descending pathways originating in the optic tectum and torus semicircularis can be further subdivided. Two nuclear areas, the nucleus magnocellularis and the nucleus laminaris have been described containing neurons with descending axons [Walkowiak and Luksch, 1994]. Within the optic tectum of *Plethodon,* one tract descending contralaterally and three tracts descending ipsilaterally could be identified as originating from four morphologically distinct types of tectal neurons [Dicke, 1998; Roth et al., 1998]. In a comparative tracer study on tectal neurons of frogs and salamanders, the same types of neurons with respect to their dendritic arborization and the pattern of ipsilaterally and bilaterally ascending (to pretectum and thalamus) and ipsilaterally or contralaterally descending projections (to nucleus isthmi, medulla oblongata and rostral spinal cord) were found [Dicke and Roth, 1996].

Intracellular staining of single neurons in the optic tectum of *Plethodon* [Roth et al., 1998] and in the torus semicircularis of *Discoglossus* [Luksch and Walkowiak, 1998] revealed that axons of tectal or toral neurons descending to the rostral spinal cord give rise to a substantial amount of axon collaterals inside the medulla oblongata. In plethodontid salamanders, application of tracers to some of the nuclei with descending axons such as the striatum, thalamic nuclei, the suprachiasmatic nucleus, the red nucleus and the optic tectum confirms the findings that axons form many collaterals remaining in the vicinity of the descending tract inside the medulla oblongata even when axons further descend to the spinal cord [Dicke, 1998; Dicke, unpubl. data]. Therefore, it is reasonable to assume that the nuclei constituting the descending pathways to the medulla oblongata and spinal cord massively influence the reticular network of the medulla oblongata. The projection nuclei of the various brain parts giving rise to the descending pathways to the reticular formation are assumed to be involved in a variety of tasks such as motivation, emotion, attention, vegetative and sensory functions contributing to the control of feeding behavior.

Despite great similarities in the overall distribution of descending pathways and their cell groups of origin, the following minor differences were found between *Discoglossus* and *Plethodon.* (1) In one specimen of *Plethodon,* but not in *Discoglossus,* a few labeled neurons were found in the transition zone between the lateral pallium and striatum. (2) In salamanders, only few neurons were found in the rostro-dorsal thalamus after ventromedial tracer application, while in *Discoglossus* a substantial number of labeled neurons was found. (3) In salamanders, neurons in the nucleus of the fasciculus longitudinalis medialis were labeled even after lateral tracer application, which was not the case in *Discoglossus.* (4) In salamanders, neurons in the ipsilateral tectum were labeled after ventromedial and ventrolateral tracer application, whereas in *Discoglossus* they were labeled only after ventrolateral tracer application. An explanation for findings 1 and 2 is that the presence and number of labeled neurons are highly dependent on the success of tracer application, which is not fully controllable. Findings 3 and 4 may be explained by the fact that in salamanders the medulla oblongata is narrower compared to frogs and that descending pathways do overlap to a greater extent, e.g. with their collaterals.

Electrophysiological Properties of Tectobulbar and Tectospinal Pathways

Our electrophysiological recordings in *H. italicus* demonstrate that the ipsilaterally and contralaterally descending tectobulbospinal pathways mostly act in an excitatory way on motor neurons of the hypoglossal nerve via mono-, oligo- and polysynaptic contacts. Only few inhibitory responses, in addition to excitatory-inhibitory ones, were found. In plethodontid salamanders, the hypoglossal nucleus, together with the 2nd spinal nucleus, controls – among other muscles – the tongue retractor musculus rectus cervicis profundus. Responses of the hypoglossal nerve, i.e. of axons of hypoglossal motor neurons, occur with minimal latency of 5.2 and 5.6 ms at ipsilateral and contralateral optic tectum stimulation, respectively, and precisely follow repetitive stimulation at very short intervals. The same was the case with responses at intracellular recordings from hypoglossal motor neurons. When stimulus intensity was increased, response amplitudes were larger and peak latencies were longer, but the onset of latencies was not shortened. This suggests that the initial part of the excitatory response of the motor neurons is mediated by mono- or oligosynaptic tectal input. However, at least in some hypoglossal motor neurons there were clear signs of spatial and temporal summation of the EPSPs, e.g. after combined application of shocks to the ipsi- and contralateral optic tectum. This strongly suggests that polysynaptic connections mediated by reticular interneurons are involved in the responses as well. Extracellular recordings from units in the white matter ventral to hypoglossal motor neurons revealed responses with latencies almost identical to that of the earliest onset of latencies of hypoglossal motor neurons; the responses, again, followed precisely repetitive stimulation of the ipsilateral and contralateral optic tectum at short intervals. These results suggest that units in the white matter ventral to the hypoglossal nucleus are probably axons of reticular interneurons that receive mono- or oligosynaptic input from the tectum and are presynaptic to the hypoglossal motor neurons. This assumption is further corroborated by the responses of both premotor candidates and motor neurons to tetanic stimulation of the optic nerve. This tetanic stimulation apparently enhances the response of hypoglossal motor neurons through short-lasting enhancement of excitability in both premotor interneurons and motor neurons. The existence of monosynaptic input to hypoglossal motor neurons from the ipsilateral and contralateral tectum in addition to oligo- and polysynaptic input mediated by reticular interneurons has gained additional support from double-labeling experiments under light-microscopical conditions [Dicke, 1992]; however, this has to be confirmed electron-microscopically.

Functional Considerations

The electrophysiological experiments indicate striking differences between frogs, as revealed by experiments in the Japanese toad, *B. japonicus* [Matsushima et al., 1985, 1986] and salamanders in the neuronal pathways from the optic tectum to hypoglossal motoneurons. First, the synaptic actions of the optic tectum are mainly excitatory in the salamander, while they are composed of early inhibition followed by excitation in the toad. Second, the major excitatory actions are mediated by oligosynaptic (and possibly monosynaptic) pathways in salamanders, while those in the toad are mediated by polysynaptic pathways. Third, the premotor candidates in the white matter ventral to the hypoglossal motor nucleus, presumably the descending axons of medullary reticular neurons, are most probably excitatory interneurons in the salamander [Dicke and Roth, 1994; Heimbuch and Dicke, 1998], whereas they are most likely inhibitory interneurons in the toad [Matsushima et al., 1989]. Finally, in the salamander, the excitatory actions showed strong enhancement suggesting that tectal descending activities are temporally integrated at the level of reticular interneurons and motor neurons.

Thus, the tectospinal neural pathways involved in feeding seem to be more direct in salamanders and – at least in the case of tongue-projecting salamanders – much faster than in frogs. Bolitoglossine salamanders such as *Bolitoglossa occidentalis* and *H. italicus* are capable of fully extruding their tongue within 5–10 ms [Roth, 1976; Thexton et al., 1977]. Whether these differences are due to a fundamental difference in the temporal sequence of muscular activities during feeding behavior is presently under investigation in our laboratory. While our knowledge about the functional organization of premotor and motor centers in the amphibian medulla oblongata and rostral spinal cord is still scarce, we can nevertheless state the following. Neither in frogs/toads nor in salamanders is there any sign that tectal visual neurons directly control feeding behavior, despite the existence of a few monosynaptic connections of descending tectal fibers with motor neurons at least in salamanders. Rather, both in anurans and urodeles, there are at least four separate tectobulbospinal pathways that terminate in the reticular formation surrounding the motor nuclei related to feeding or give off collaterals to them [Ten Donkelaar et al., 1981; Ten Donkelaar, 1982; Tóth et al., 1985, for frogs; Dicke and Roth, 1994; Dicke, 1998; Roth et al., 1998, for salamanders].

Neurophysiological studies on reticular neurons in the toad *Bufo bufo* [Ewert et al., 1990; Schwippert et al., 1990] revealed highly complex response properties at visual stimulation with prey-like stimuli. While some classes of reticular neurons faintly resembled response types of tectal neurons, others showed completely different responses. Generally, the receptive fields of visually driven reticular neurons were much larger than tectal ones and revealed no retinotopy; many of them were multimodal. A similarly complex picture was obtained from intracellular recordings from reticular neurons in the salamanders *H. italicus* and *P. jordani* at electrical stimulation of the optic tectum and other visual and nonvisual brain centers and afferent nerves [Heimbuch and Dicke, 1998]. These data suggest that the amphibian reticular formation integrates in a highly complex fashion visual and nonvisual information via pathways originating from tectal and nontectal centers described in this article. The neurophysiological results presented in this article further indicate differences between frogs/toads and salamanders in the way in which this integration takes place in the reticular formation.

In summary, our data suggest that differences in feeding behavior existing in frogs and salamanders do not result from additional input or different organization of pathways to the premotor network in the medulla oblongata. Rather, the same organization of descending pathways to the medulla oblongata in frogs and salamanders seems to represent a basic requirement for the performance of a motor action. Specific neuronal network properties that may explain the observed differences in the pattern of activating the tongue motor system and the feeding behavior are not detectable at this morphological level, but – for example – at the level of synaptic interaction among neurons (e.g. reticular neurons and motor neurons). Immunohistochemical investigations on transmitters and modulators in amphibians have shown that, for example, the dopaminergic system in the basal ganglio of frogs and salamanders is quite similar [Marín et al., 1997a], but that there are differences with respect to the cholinergic and serotonergic systems between frogs and salamanders as well as among different groups of salamanders. Cholinergic fibers have been found in the optic tectum of the two frog species *Rana perezi* and *Xenopus laevis* and the salamander *Pleurodeles waltl,* whereas in the tectum of plethodontid salamanders, e.g., *Plethodon* and *Hydromantes,* the cholinergic fiber system is present and in addition cholinergic neurons are found [Marín et al., 1997b; Wallstein and Dicke, 1998]. In *Rana pipiens,* serotonergic fibers display a laminar organization in the optic tectum and serotonergic cells are found in different cellular layers [Debski et al., 1995; Liu and Debski, 1995]. In plethodontid salamanders, a well-developed laminated fiber system is present, while in *P. waltl* only diffuse 5-HT fiber projections are observed [Dicke et al., 1997]. Thus, the observed differences may be attributed to differences in the presence and action of cholinergic and serotonergic neuromodulatory systems in the optic tectum and – possibly – the reticular formation. Studies about the cytoarchitecture, neurochemical and functional organization of the reticular formation of amphibians going on in our laboratory will further clarify this question.

Acknowledgment

This study was supported by grants from the Deutsche Forschungsgemeinschaft to UD (SFB 517) and GR (Ro 481/10-1), and by a research fellowship from the Alexander von Humboldt foundation of Germany to TM.

References

Deban, S., D.B. Wake, G. Roth (1997) Salamander with a ballistic tongue. Nature *389:* 27–28.

Debski, E.A., Q. Liu, A. Champman (1995) Nonuniform distribution of cellular phenotypes in the optic tectum of the leopard frog. J Comp Neurol *360:* 671–684.

Dicke, U. (1992) Neuroanatomische Untersuchungen zum Aufbau und zur Entwicklung der cervicospinalen Motonuclei und ihrer Verknüpfung mit den visuellen Zentren bei Salamandern; Dissertation, University of Bremen.

Dicke, U. (1997) Origins of descending projections to the rostral medulla oblongata in frogs and salamanders; in Elsner N, H Wässle (eds): Göttingen Neurobiology Report 1997. 25th Conf Neurobiol Göttingen. Stuttgart, Thieme, vol 2, p 697.

Dicke, U. (1998) Morphology, axonal projection pattern and response types of tectal neurons in plethodontid salamanders. I. A tracer study of projection neurons and their pathways. J Comp Neurol, in press.

Dicke, U., G. Roth (1994) Tectal activation of premotor and motor networks during feeding in salamanders. Eur J Morphol *32:* 106–116.

Dicke, U., G. Roth (1996) Similarities and differences in the cytoarchitecture of the tectum of frogs and salamanders. Acta Biol Hung *47:* 41–59.

Dicke, U., M. Wallstein, G. Roth (1997) 5-HT-like immunoreactivity in the brains of plethodontid and salamandrid salamanders *(Hydromantes italicus, Hydromantes genei, Plethodon jordani, Desmognathus ochrophaeus, Pleurodeles waltl):* An immunohistochemical and biocytin double-labelling study. Cell Tissue Res *287:* 513–523.

Duellman, W.E., L. Trueb (1986) Biology of Amphibians. New York, McGraw-Hill.

Ewert, J.P. (1974) The neural basis of visually guided behavior. Sci Am *230:* 34–42.

Ewert, J.P. (1984) Tectal mechanisms that underlie prey-catching and avoidance behaviors in toads, in Vanegas H (ed): Comparative Neurology of the Optic Tectum. New York, Plenum Press, pp 247–416.

Ewert, J.P. (1989) The release of visual behavior in toads: Stages of parallel/hierarchical information processing; in Ewert JP, M Arbib (eds): Visuomotor coordination: Amphibians, Comparisons, Models, and Robots. New York, Plenum Press, pp 39–120.

Ewert, J.P., E.M. Framing, E. Schürgg-Pfeiffer, A. Weerasuriya (1990) Responses of medullary neurons to moving stimuli in the common toad. I. Characterization of medial reticular neurons by extracellular recording. J Comp Physiol *167:* 495–508.

Heimbuch, J., U. Dicke (1998) Neurons of the medial reticular zone in lungless salamanders: Physiology and morphology; in Elsner N, R Wehner (eds): Göttingen Neurobiology Report 1998. Proc 26th Conf Neurobiol Göttingen, Thieme, Stuttgart, vol 2, p 499.

Herrick, C.J. (1948) The Brain of the Tiger Salamander *Ambystoma tigrinum.* Chicago, University of Chicago Press.

Liu, Q., E.A. Debski (1995) Origins of serotonin-like immunoreactivity in the optic tectum of *Rana pipiens.* J Comp Neurol *352:* 280–296.

Luksch, H., W. Walkowiak (1998) Morphology and axonal projection patterns of auditory neurons in the midbrain of the painted frog, *Discoglossus pictus.* Hear Res *122:* 1–17.

Marín, O., W.J.A.A. Smeets, A. González (1997a) Basal ganglia organization in amphibians: Catecholaminergic innervation of the striatum and the nucleus accumbens. J Comp Neurol *378:* 50–69.

Marín, O., W.J.A.A. Smeets, A. González (1997b) Distribution of choline acetyltransferase immunoreactivity in the brain of anuran *(Rana perezi, Xenopus laevis)* and urodele *(Pleurodeles waltl)* amphibians. J Comp Neurol *382:* 499–534.

Matsushima, T., M. Satou, K. Ueda (1985) An electromyographic analysis of electrically-evoked prey-catching behavior by means of stimuli applied to the optic tectum in the Japanese toad. Neurosci Res *3:* 154–161.

Matsushima, T., M. Satou, K. Ueda (1986) Glossopharyngeal and tectal influences on tongue-muscle motoneurons in the Japanese toad. Brain Res *365:* 198–203.

Matsushima, T., M. Satou, K. Ueda (1989) Medullary reticular neurons in the Japanese toad: Morphologies and excitatory inputs from the optic tectum. J Comp Physiol *166:* 7–22.

Naujoks-Manteuffel, C., G. Manteuffel (1988) Origins of descending projections to the medulla oblongata and rostral medulla spinalis in the urodele *Salamandra salamandra* (Amphibia). J Comp Neurol *273:* 187–206.

Naujoks-Manteuffel, C., G. Manteuffel, W. Himstedt (1988) On the presence of nucleus ruber in the urodele *Salamandra salamandra* and the caecilian *Ichthyophis kohtaoensis.* Behav Brain Res *28:* 29–32.

Neary, T.J., R.G. Northcutt (1983) Nuclear organization of the bullfrog diencephalon. J Comp Neurol *213:* 262–278.

Nieuwenhuys, R., P. Opdam (1976) Structure of the brain stem; in Llinas R, W Precht (eds): Frog Biology. Berlin, Springer, pp 811–855.

Opdam, P., R. Nieuwenhuys (1976) Topological analysis of the brain stem of the axolotl *Ambystoma mexicanum.* J Comp Neurol *165:* 285–306.

Puelles, L., F.J. Milán, M. Martínez de la Torre (1996) A segmental map of architectonic subdivisions in the diencephalon of the frog *Rana perezi:* Acetylcholinesterase-histochemical observations. Brain Behav Evol *47:* 279–310.

Roth, G. (1976) Experimental analysis of prey catching behavior of *Hydromantes italicus* Dunn (Amphibia, Plethodontidae). J Comp Physiol *109:* 47–58.

Roth, G.: Visual Behavior in Salamanders. Berlin, Springer, 1987.

Roth, G., K.C. Nishikawa, U. Dicke, D.B. Wake (1988) Topography and cytoarchitecture of the motor nuclei in the brainstem of salamanders. J Comp Neurol *278:* 181–194.

Roth, G., K.C. Nishikawa, D.B. Wake, U. Dicke, T. Matsushima (1990) Mechanics and neuromorphology of feeding in amphibians. Neth J Zool *40:* 115–135.

Roth, G., U. Dicke, W. Grunwald (1998) Morphology, axonal projection pattern and response types of tectal neurons in plethodontid salamanders. II. Intracellular recording and labeling experiments. J Comp Neurol, in press.

Satou, M., J.P. Ewert (1985) The antidromic activation of tectal neurons by electrical stimuli applied to the caudal medulla oblongata in the toad *Bufo bufo* L. J Comp Physiol *157:* 739–748.

Satou, M., T. Matsushima, H. Takeuchi, K. Ueda (1985) Tongue-muscle controlling neurons in the Japanese toad, *Bufo japonicus:* Topography, morphology, and neuronal pathways for their activation and inhibition by descending volleys from the optic tectum. J Comp Physiol *157:* 717–737.

Schwippert, W.W., T.W. Beneke, J.P. Ewert (1990) Responses of medullary neurons to moving stimuli in the common toad. II. An intracellular recording and cobalt-lysine labeling study. J Comp Physiol *167:* 509–520.

Székely, G., G. Lázár (1976) Cellular and synaptic architecture of the optic tectum; in Llinas R, W Precht (eds): Frog Neurobiology. Berlin, Springer, pp 407–434.

Ten Donkelaar, H.J. (1982) Organization of descending pathways to the spinal cord in amphibians and reptiles; in Kuypers HGJM, GF Martin (eds): Progress in Brain Research. Amsterdam, Elsevier, vol 57, pp 26–58.

Ten Donkelaar, H.J., R. de Boer-van Juizen, F.T.M. Schouten, S.J.H. Eggen (1981) Cells of origin of descending pathways to the spinal cord in the clawed toad *(Xenopus laevis).* Neuroscience *6:* 2297–2312.

Thexton, A.J., D.B. Wake, M.H. Wake (1977) Control of tongue function in the salamander *Bolitoglossa subpalmata.* Arch Oral Biol *22:* 361–366.

Tóth, P., G. Csank, G. Lázár (1985) Morphology of the cells of origin of descending pathways to the spinal cord in *Rana esculenta.* A tracing study using cobaltic-lysine complex. J Hirnforsch *26:* 365–383.

Wake, D.B., K.C. Nishikawa, U. Dicke, G. Roth (1988) Organization of the motor nuclei in the cervical spinal cord of salamanders. J Comp Neurol *278:* 195–208.

Walkowiak, W., H. Luksch (1994) Sensory motor interfacing in acoustic behavior of anurans. Am Zool *34:* 685–695.

Wallstein, M., U. Dicke (1998) Choline acetyltransferase-like immunoreactivity in the optic tectum of the plethodontid salamander *Plethodon jordani;* in Elsner N, R Wehner (eds): Göttingen Neurobiology Report 1998. Proc 26th Conf Neurobiol Göttingen. Stuttgart, Thieme vol 2, p 453.

Acta Anatomica

Acta Anat 1998;163:144–156

Received: April 9, 1998
Accepted: October 21, 1998

R.G. Bout

Evolutionary Morphology,
Institute of Evolutionary and
Ecological Sciences, Leiden University,
Leiden, The Netherlands

Complex Movement Patterns: Modifiability and Constraints

Key Words
Motor pattern
Kinematics
Feeding
Drinking
Plasticity

Abstract
Most behaviours involve complex morphological systems and vice versa morphological systems are used by the organism in many different ways. During evolution and ontogeny changes in kinematics and function of skeletal and muscular systems must be coordinated with changes in their neural control. Neuromotor patterns are sometimes believed to be conserved in evolution, leading to diversification at the level of musculoskeletal design. Vertebrate motor patterns used in feeding are reviewed to examine this hypothesis. Stereotyped behaviour is not necessarily the result of phylogenetic constraints but may also result from the functional demands imposed by the mechanics of the jaw apparatus and the nature of the task performed. Sensory feedback and descending control not only contribute to 'online' control of movement but also shape the development of motor patterns and learning behaviour and indicate a potentially large flexibility. The neural and sensory apparatus that produces this flexibility will be subject to evolutionary modification. In the absence of a demand for flexibility motor patterns may become stereotyped in some species, while they are very flexible in others. To the extent that morphological systems perform independent movements during different behaviours, separate basic motor patterns may be required, which may be coordinated in different ways.

Introduction

One of the key problems in the study of evolution is: how do complex structures evolve? Most behaviours involve complex morphological systems, composed of many different parts, and vice versa morphological systems are used by the organism in many different ways. During evolution and ontogeny changes in structurally and functionally highly integrated systems may easily compromise (one of) the different functions performed by the system. Kinematics and function of skeletal and muscular systems, for instance, must be coordinated with changes in their neural control. Transformations in movement patterns may be produced by changes in the musculoskeletal system, changes in motor pattern, or both (fig. 1). The question of how to change an integrated system without compromising its performance addresses on the one hand the problem of flexi-

Abbreviations used in this paper:
AME(M) = Musculus adductor mandibulae externus (pars medialis); CPG = central pattern generator; DM = musculus depressor mandibulae; FC = fast closing phase; FO = fast opening phase; PQP = musculus protractor quadrati et pterygoidei; PTP = musculus pseudotemporalis profundus; PVL = musculus pterygoideus pars ventralis lateralis; SC = slow closing phase; SO = slow opening phase.

KARGER

Fax + 41 61 306 12 34
E-Mail karger@karger.ch
www.karger.com

0001–5180/98/1633–0144$17.50/0

Accessible online at:
http://BioMedNet.com/karger

Dr. R.G. Bout
Neurobehavioral Morphology, Institute of Evolutionary and Ecological Sciences
Van der Klaauw Laboratory, Kaiserstraat 63
POB 9516, NL–2300 RA Leiden (The Netherlands)
Tel. +31 71 5274945, Fax +31 71 5274900, E-Mail bout@rulsfb.leidenuniv.nl

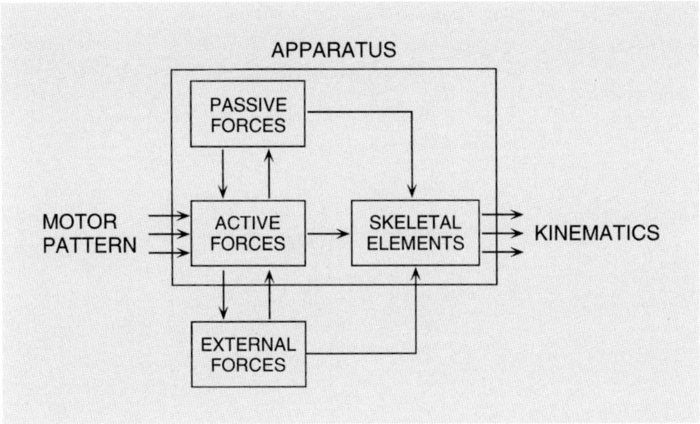

Fig. 1. Schematic diagram illustrating the interaction of passively and actively generated forces and the problem of infering motor patterns from kinematics.

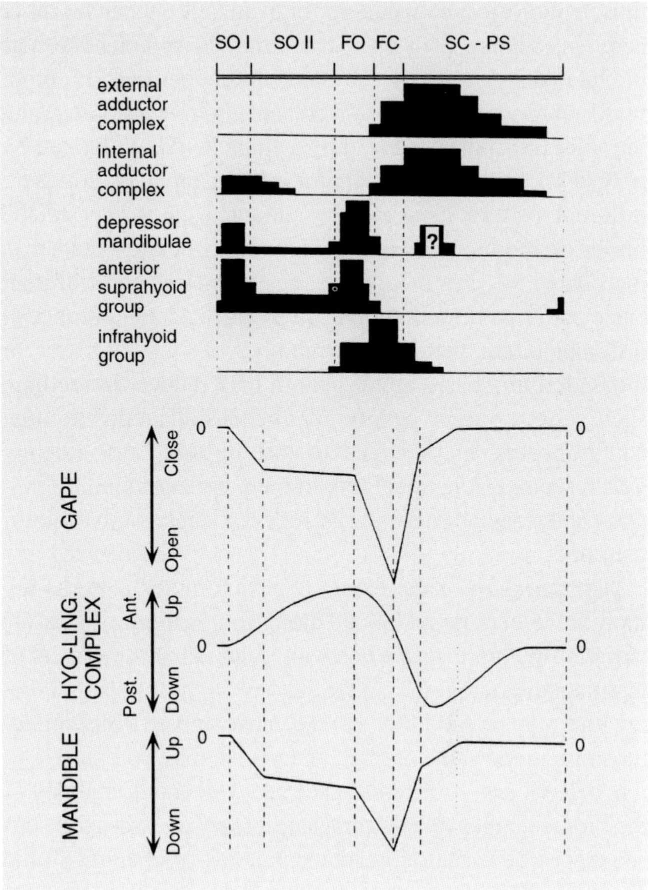

Fig. 2. Generalized tetrapod intraoral transport cycle [from Bramble and Wake, 1985].

bility of characters in morphological systems in general, on the other hand there is the question whether some characters or parts of a system are more suited to accomplish changes than others.

Neuromotor patterns are sometimes believed to be conserved in evolution [Lauder, 1991]. It has also been suggested that constraints on the level of neuromotor patterns have led to diversification at the level of musculoskeletal design [Roth and Wake, 1989]. In feeding systems one of the reasons that the hypothesis of preprogrammed patterns has been readily adopted is that a central pattern generator (CPG) may produce rhythmic opening and closing. Furthermore, similarities in kinematics among different taxa are used to infer an underlying general pattern. However, the mechanics of the feeding apparatus among vertebrates are very different. Movement patterns are the expression of all actively generated muscle forces (motor pattern), passive forces in the musculoskeletal system (the feeding apparatus), and external forces (e.g., food reaction forces) acting on the system. The interaction of these types of forces suggests that similar kinematics do not necessarily imply similar neuromotor patterns and a change in kinematics does not always have to coincide with a change in motor pattern. The problem with the idea of a conservative general motor pattern is that the specific features that would identify such a pattern, as opposed to a pattern determined by convergent functional requirements, have never been defined [but see Smith, 1994].

A neuromotor pattern, or simply motor pattern, of feeding is measured as the electrical activity (EMG) of jaw muscles and is thought to reflect the activity of the putative CPG. However, the motor patterns are variable and modifiable by peripheral input and descending central control mechanisms. The 'basic' neuromotor pattern is often distinguished from feedback modulation by separating effects on the sequence and relative onset of muscle activity from effects on amplitude and duration. Such an approach ignores variation in motor patterns and generalizes a 'model' feeding cycle that is often limited to a single behaviour, e.g., intraoral transport cycles (fig. 2). However, the anatomical analysis of the neural substrate suggests that the various kinematic elements of a feeding cycle share not only the same motor system, but to a large extent also the same premotor system [see the accompanying paper by Dubbeldam, 1998]. When all parts of a feeding cycle are analyzed, motor patterns for grasping, repositioning, reduction, transport and ingestion are similar in many respects, although they may all show their own characteristics. When such motor patterns are compared, does the absence of activity in one jaw muscle or an extra burst of activity in another muscle indicate a new pattern (or even CPG) when all other activities remain the same, or are they to be considered as modulated versions of the same basic pattern? In other words are motor patterns for prehension, repositioning, and intraoral transport produced by a single CPG modulated by peripheral sensory input and/or central input or are there different pattern generators? Without a clear idea of the limits of modulation, one cannot discriminate between a modulated version of a single basic pattern, and different basic patterns. If modulation can produce a large number of different motor patterns, the number of characteristics of the underlying basic motor pattern may reduce. In feeding, such a basic motor pattern for prehension, repositioning, and transport, may reduce to just opening and closing, which seems a functional requirement for most feeding systems and does not necessarily reflect a conservative motor pattern.

The question of the extent to which motor patterns are modifiable may be addressed along four different lines: (1) variation within a single behaviour, including the effects of learning and ontogeny, (2) the sensory input for a single behaviour may be modified, (3) we may compare different behaviours in the same animal, and (4) we may compare similar behaviours in different species. A detailed analysis of the biomechanics of a system is necessary to assess the consequences of variation in motor patterns or changes in the musculoskeletal system. However, there are very few examples, even for relatively simple behaviours such as eating and drinking, for which both the biomechanics and the motor patterns involved are completely known.

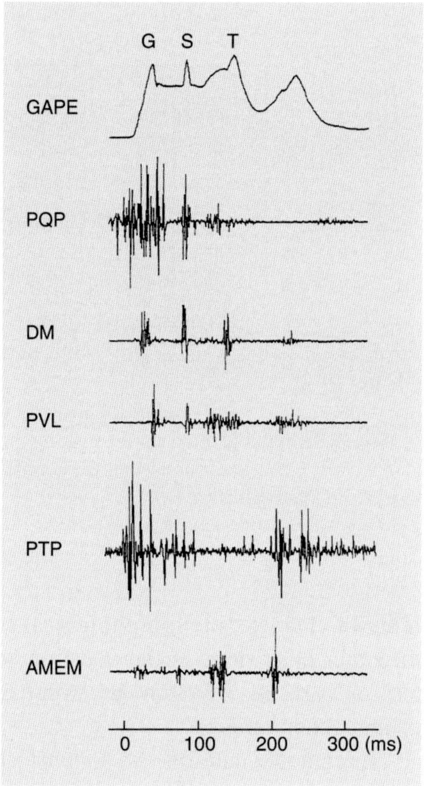

Fig. 3. Gape (top trace) during grasping (G), a stationing movement (S) and an intraoral transport cycle (T) of a food pellet by a pigeon. Upward and downward movements indicate jaw opening and jaw closing. The jaw muscle activities are recorded from the upper jaw opener/quadrate protractor (PQP), lower jaw opener (DM), a pterygoid muscle (PVL), a quadrate retractor (PTP) and a lower jaw adductor (AMEM).

Jaw Mechanics of the Pigeon: A Case Study

As an example of the complex biomechanics of the jaw apparatus we have analyzed eating and drinking in pigeons. Eating in pigeons involves three distinct behaviours: grasping, stationing and intraoral transport. Grasping involves jaw opening and terminates when the seed is held between the beaks; stationing involves rapid opening and closing movements of the beak, which serve to reposition the seed prior to transport; intraoral transport involves a lingual transport mechanism, an inertial transport mechanism or a combination of both (fig. 3). During intraoral transport there is a slow opening phase prior to the fast opening phase. During this slow phase the tongue is protruded and pushed against the seed. Slow opening is absent during grasping or stationing.

The biomechanical apparatus that produces these different behaviours in birds is characterized by (1) a kinematic upper jaw that hinges around a flexible zone and (2) the presence of a ligament originating on the skull and inserting to the lower jaw in front of the quadratomandibular joint. In some birds this postorbital ligament is absent, e.g., finches [Classen, 1989; Nuijens and Zweers, 1997; for other species see Fiedler, 1951; Goodman and Fisher, 1962; Burton, 1974], or connects to the jugal bar rather than the mandible, e.g. paleognathous birds [Hofer, 1950; Fiedler, 1951; Elzanowski, 1987]. The movements of the pigeon's beak are mediated by many different muscles, which fall into four functional groups: the depressors of the lower jaw, the adductors connecting lower jaw and skull, the protractors of the quadrate and pterygoid, and the pterygoid muscles. For opening and closing of the jaws three points are of special interest. First, none of the muscles insert directly on the upper jaw. Opening of the upper jaw is produced indirectly by the protractor muscles operating on the pterygoid and quadrate: by pulling this chain of bones forward the maxilla is elevated. Note, however, that the protractor muscles are not the only muscles that are able to pull the quadrate forward. Contraction of adductors and depressors may also lift the quadrate and assist in opening the upper jaw [Van Gennip and Berkhoudt, 1992]. Second, the cranium and mandible are connected by a ligament which spans the quadratomandibular and quadratocranial joint. The stretch in collagenous ligaments is limited (6–10%) and the insertion of the ligament in front of the quadratomandibular joint is sometimes believed to effectively lock the lower jaw [see Bock, 1964]. As the distance between the quadratomandibular joint and the insertion of the ligament is very small, a considerable gape may be realized before the ligament is stretched to its limit. In the zebra finch depressor force may be sufficient to lower the mandible 6 mm against the resisting force of the stretched postorbital ligament [Nuijens and Bout, 1998]. The third point of interest is that there are two different ways to prevent stretching the ligament and to lock the lower jaw. First, quadrate rotation will protract the quadrate-pterygoid-palatine complex and produce not only opening of the upper beak, but also slackening of the postorbital ligament, which unlocks the lower jaw. Alternatively, Van Gennip and Berkhoudt [1992] have demonstrated in a three-dimensional kinematic model of the pigeon skull that slackening of the ligament may be produced by levating the quadrate without moving it forward. In this second mode of jaw opening muscle force is used to keep the upper jaw as its resting position and to change the angle between pterygoid and quadrate. This also produces slackening of the ligament. In this position the lower jaw is 'unlocked' without opening the upper beak. The presence of two jaw opening mechanisms suggests that pigeons may use different motor patterns to produce a gape.

Movement [Klein et al., 1985] and EMG analyses [Bout and Zeigler, 1994a] provide experimental evidence for the kinematic hypotheses discussed. The main differences in the motor patterns for grasping, stationing and intraoral transport are the presence or absence of an upper jaw opener burst, and the presence of adductor muscle activity (fig. 4). In pecking pigeons upper beak elevation during grasping always precedes lower jaw opening. This is consistent with EMG analysis (fig. 3): the onset of jaw opening during grasping is preceded by activity of the jaw elevator muscle (musculus protractor quadrati et pterygoidei, PQP), followed after 10 ms by activity of the lower jaw opener. The activity of jaw opener muscles, which act against passive forces, is relatively constant during grasping even when sensory input (pellet size) varies. However, during repositioning of a seed the upper beak is only clearly elevated during the first stationing, but elevation is much less or even absent during second or third stationings [unpubl. data; see also Van Gennip, 1988]. The activity of the upper jaw opener muscles follows the same pattern: activity is present during first stationing movements, but often absent in following stationing movements. During intraoral transport upper jaw opener activity is present during ingestion of small seeds, but absent in large seeds. In those instances where the bird does not elevate its upper beak it is apparently using the second mode of jaw opening.

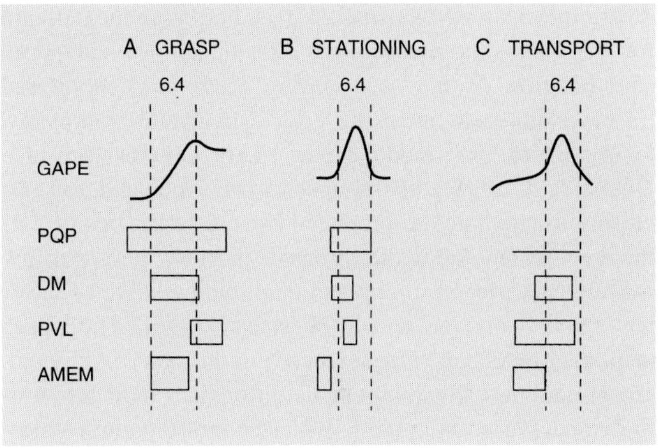

Fig. 4. Schematic diagram illustrating the onset and offset of jaw muscle activities for grasping, stationing and intraoral transport during the ingestion of a 6.4-mm food pellet. The dashed vertical lines represent the start of (fast) opening and the end of closing.

The role of this rather complicated biomechanical mechanism may be to provide the bird with a 'snap-shut' mechanism. During grasping, jaw closing is clearly related to the onset of pterygoid muscle activity and not to adductor activity, which more or less coincides with jaw muscle opener activity and decreases before jaw closing starts. The velocity of head descent towards the seed is very high [500–1,000 mm/s; see Klein et al., 1985] and the direction of movement is reversed almost instantaneously after reaching the ground. This leaves a very short time (5–10 ms) to get a firm grip on the seed. The elastic energy that is stored in the prokinetic joint of the upper beak is released when the pterygoid muscles become active and start to act as antagonists of the upper jaw openers. The elastic forces will immediately close the jaws. A similar situation may exist when head jerks are used in stationing and intraoral transport, but the use of sticky saliva to 'glue' a (small) seed to the tongue or a more caudal position of the seed may decrease the chance of loosing the seed and relax the demand for the snap-shut mechanism, e.g., upper jaw elevation. The position of the seed during feeding is monitored by mechanoreceptors in the beak [Silver and Witkovsky, 1973], which may modulate the motor patterns [see also Dubbeldam, 1998]. In the absence of a demand for fast closing a gape may be produced by a different position of the jaws. In singing white-throated sparrows the upper jaw movement varies in timing and amplitude relative to lower jaw movement [Hoese and Westneat, 1996]. The upper beak may be elevated or depressed in the same individual, repeating a note. Such kinematic differences indicate intra-individual variation in the jaw muscle motor pattern. Interestingly, the upper beak is not elevated during grasping in *Rhea* [Gussekloo and Zweers, 1997].

The mechanical and EMG analysis in this case study shows that in the pigeon there are two jaw opening mechanisms, typically involving the presence or absence of activity of the upper jaw opener muscles. These two motor patterns are not just used for different parts of the feeding cycle, but also for a single phase in the feeding cycle (stationing, intraoral transport). The absence of such variation in grasping seems to follow from the functional demand for high velocity closing. Furthermore, all characteristics of a motor pattern (sequence, relative onset, amplitude and duration of activity) vary, though not all in the same motor pattern.

Motor and Movement Patterns in Other Vertebrates

Unfortunately there are very few detailed EMG and biomechanical studies on birds. Although the basic organization of the jaw apparatus is the same in most species of birds, relatively small differences in morphology may result in different kinematics of the jaw system. The motor pattern of grasping in chicken is very similar to that of the pigeon, although, unlike in the pigeon, adductor activity is related to the onset of closing. The general similarity is hardly surprising, as the basic design of the feeding apparatus, and the functional demands imposed by pecking seeds are the same. A detailed analysis of upper and lower jaw movements in the chicken suggests movement stereotypy for grasping [Van den Heuvel and Berkhoudt, 1998], although pterygoid and adductor muscles do vary in their time of onset [Van der Heuvel, 1992] for the same seed size. The lack of EMG data in birds makes it impossible to properly evaluate conservation of feeding motor patterns. As indicated in the introduction movement patterns alone are insufficient to infer neuromotor patterns in any detail. As illustrated by the biomechanics of the pigeon head, the interaction of muscle forces (motor pattern), passive forces, and external forces is complex and similar kinematics do not necessarily imply similar neuromotor patterns or vice versa.

In reptiles, a variety of motor patterns is present, which seem related to differences in skull mechanics and movement. Although many reptiles have a (meso) kinetic skull, in most species movement of the quadrate seems independent of skull movement and the muzzle is not lifted during jaw opening [Herrel, pers. commun.]. A comparison of motor patterns among reptiles and birds shows some unexpected instances of coactivation of adductor and depressor muscles. During intraoral transport in snakes there is no fast opening and a clear (unilateral) alternation of closer and depressor activity [Kardong and Berkhoudt, 1998]. Interestingly, the presence of a mobile upper jaw is, as in birds, associated with pterygoid activity during the early phase of closing, which is at odds with the general tetrapod feeding model. In *Agama stellio* [Herrel et al., 1997] the fast opening and closing phases of intraoral transport show a clear alternation of high amplitude depressor and adductor activity, although there may be some low level activity in the antagonist in both phases. As in the pigeon there is low level adductor activity, while the tongue protrudes during the slow opening phase. In *Alligator mississipiensis* the skull is akinetic and there is no movement of the quadrate [Busbey, 1989]. Food processing is minimal and the general sequence of food reduction and intraoral transport is by inertial movements of the head. During jaw opening, which is sometimes interrupted, only the depressor muscle is active. However, closing movements are fast and involve clear coactivation of adductors and depressors. Coactivation of depressors and adductors during jaw closing is absent in *Sphenodon* [Gorniak et al., 1982], *Trachydosaurus* [Gans et al., 1985], and *Varanus exanthematicus* [Smith, 1982]. However, coactivation is sometimes found during jaw opening. The monitor lizard uses inertial transport of food to ingest large prey [Smith, 1982]. During the opening phase of intraoral transport there is coactivation of (part of the) adductor and pterygoid muscles with the depressor muscles. A somewhat similar situation is found in the pigeon (fig. 4). Adductors are active during jaw opening in grasping but not during jaw opening or closing in stationing and show low level activity throughout intraoral transport cycles.

Similar differences in coactivation may exist for crushing phases. In reptiles the depressor muscle is not active during the crushing phase [Herrel et al., 1997; Busbey,

Acta Anatomica

Acta Anat 1998;163:157–172

Received: March 9, 1998
Accepted: August 24, 1998

J.L. Dubbeldam

Institute of Evolutionary and Ecological Sciences, Leiden University, Leiden, The Netherlands

The Neural Substrate for 'Learned' and 'Nonlearned' Activities in Birds: A Discussion of the Organization of Bulbar Reticular Premotor Systems with Side-Lights on the Mammalian Situation

Key Words
Premotor system
Fleeding
Locomotion
Respiration
Vocalization

Abstract

The reticular formation of the brainstem contains premotor systems for various musculomotor systems. In this paper, the bulbar premotor systems for jaw and tongue movements, head and neck movements, locomotion, and respiration and vocalization in birds are reviewed and compared to premotor systems in mammals. Roughly, the bulbar reticular formation can be subdivided in three longitudinal zones: a dorsolateral (RPcdl) and a ventromedial (RPcvm) parvocellular zone and a gigantocellular zone (RGc). RPcdl contains premotor neurons for the jaw and neck system, RPcvm for the jaw, tongue and neck system, and RGc for the tongue and locomotory system. RPcdl receives input from the descending sensory trigeminal system, parts of RPcvm and RGc from vestibular nuclei, whereas the tectum has a projection to the contralateral RGc. RPcdl and RPcvm receive substantial telencephalic input through the occipitomesencephalic tract. The bulbar part of the respiratory system consists of a series of cell groups in the ventrolateral reticular formation and has connections with motor centers of the vocalization system. The similarities and differences between the avian and mammalian situation are discussed. Musculomotor systems participate in various activities. It is argued that a premotor system should possess sufficient flexibility to control the participation of a motor system in the different activities. This flexibility may permit the occurrence of learning processes in terms of refining basically existing motor patterns. The emergence of new and more complex motor patterns as in vocalization requires the involvement of hierarchically higher brain centers.

Abbreviations used in this paper:
CMOd, CMOv = Nucleus centralis medullae oblongatae, pars dorsalis, pars ventralis;
mesV = nucleus mesencephalicus nervi trigemini; RAm = nucleus retroambigualis;
RGc = nucleus reticularis gigantocellularis; RPc = parvocellular zone; RPcdl,
RPcvm = nucleus reticularis parvocellularis, zona dorsolateralis, zona ventromedialis;
vVRG = rostral ventral respiratory group.

KARGER

Fax +41 61 306 12 34
E-Mail karger@karger.ch
www.karger.com

0001–5180/98/1633–0157$17.50/0

Accessible online at:
http://BioMedNet.com/karger

Prof. Dr. J.L. Dubbeldam
Institute of Evolutionary and Ecological Sciences, RUL
POB 9516, NL–2300 RA Leiden (The Netherlands)
Tel. +31 71 5275039, Fax +31 71 5274900
E-Mail dubbeldam@rulsfb.leidenuniv.nl

Introduction

The assumed stereotypy of specific avian activities such as pecking and drinking has often been ascribed to the existence of fixed action patterns. At the same time, a certain flexibility of the movement patterns can be recognized. Bramble and Wake [1985] present a hypothetical 'model generalized feeding cycle' incorporating very similar trains of stereotyped kinematic and motor events, possibly involving homologous sets of muscles, in the various classes of tetrapod vertebrates. Such a model feeding cycle may have existed in a primitive, generalized tetrapod and consideration of the model may help to improve our understanding of the evolutionary diversification of the feeding mechanism [Bramble and Wake, 1985]. Comparative studies can help to recognize generalized patterns of motor activity during the various activities reflecting the fixed action patterns, whereas functional diversification may depend upon evolutionary changes of mechanical systems rather than upon changes of neural systems [discussions in Lauder, 1991; Nishikawa et al., 1992; Dicke and Roth, 1998]. In a companion paper Bout [1998] discusses the existence, or rather the absence of stereotypies in feeding mechanisms of several groups of vertebrates. His survey leads to the conclusion that the concept of the fixed action pattern should be replaced by the concept of modal action pattern. The latter concept implies a high degree of flexibility or plasticity of motor patterns.

The aim of the present contribution is to examine the neural substrate involved in the generation of motor patterns. Emphasis will be on brainstem (pre-)motor systems. For a better understanding of the organization of this neural substrate it is helpful to have some knowledge about the interaction of the different mechanisms during a specific action. For example, according to the model of Bramble and Wake [1985], feeding and drinking involve, at least, movements of jaws, tongue and head and neck. Then, it is necessary to consider the coherence of the motor control mechanisms of these three mechanical systems. In particular parts of the pontobulbar reticular formation form part of the neural substrate for these control mechanisms.

I shall start this overview with a consideration of the feeding system and then extend the discussion to other motor activities. Emphasis will be on the avian situation, but with reference to the corresponding systems in mammals. Relatively little is known about the organization of the reptilian premotor systems. However, birds and mammals represent parallel lines of development and comparison of corresponding systems in the two groups may provide a better understanding of the control systems in a wider sense.

The Motor System for Feeding and Drinking Movements

Feeding in birds consists of rather complex movement patterns as the opening and closing of the beak depend upon the movements of both upper and lower jaw. Four groups of muscle are responsible for these movements (fig. 1): (1) depressor muscles of the lower jaw, innervated by the dorsal facial motor nucleus, VII_d, (2) protractor quadrati muscles, innervated by cell group V_4 of the trigeminal motor complex, (3) adductor mandibulae muscles, innervated by the trigeminal cell group V_2, and (4) the pterygoid muscles, innervated by trigeminal cells group V_1 [Wild and Zeigler, 1980; den Boer et al., 1986]. Roughly, the depressor muscles are responsible for the abduction of the lower jaw, the adductor muscles for the adduction, whereas activity of the protractor muscles results in an upward movement of the upper jaw and that of the pterygoid muscles in a downward movement. Though this summary is a simplification of a very complex mechanism [e.g., Zweers, 1982a; van Gennip and Berkhoudt, 1992], it is sufficient for the present discussion. The most important point is that the depressor and protractor muscles can be considered beak opening musles, the adductor and pterygoids as beak closing muscles (fig. 1).

This system is different from that in mammals, as in the latter only the lower jaw is movable; at the same time, in many mammals the jaw can move both in the vertical and in horizontal direction, whereas movements in most birds are mainly in the dorsoventral direction. There are distinct differences in the structure of the jaw musculature, which may have consequences for the organization of the motor nuclei. The mammalian trigeminal motor nuclei do have a musculotopic organization [e.g., Matsuda et al., 1978; Mizuno et al., 1981] with a certain resemblance to the avian situation. For instance, in mammals the caudal part of musculus digastricus – a jaw opener – is innervated by the facial motor nucleus, as is the opener of the mandible in birds. An important function of the mammalian facial motor nucleus, however, is the innervation of the facial musculature, whereas the facial musculature in birds is restricted to musculus dermotemporalis and the large constrictor colli.

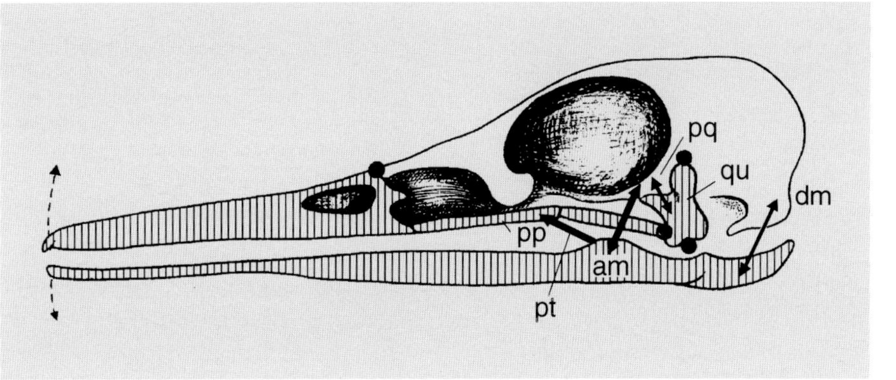

Fig. 1. Lateral schematic view of the skull of the mallard; the four main muscle groups are indicated by lines; jaw openers: dm = musculus depressor mandibulae; pq = musculus protractor quadrati; jaw closers: am = musculi adductores mandibulae; pt = musculi pterygoidei. qu = Quadrate bone; pp = pterygoid and palatine bones. Black dots indicate articulations/hinges enabling the movements of upper and lower jaw; arrows indicate direction of movement of upper and lower beak [modified after Dubbeldam, 1985].

Organization of the Bulbar Premotor System of Jaw and Tongue Movements

Tracing studies in the mallard [Bout and Dubbeldam, 1994] reveal that the parvocellular reticular formation of the lower brainstem contains premotor neurons of these motor nuclei. In Nissl-stained preparations of the avian brainstem, it is possible to recognize a lateral parvocellular zone (RPc) and a ventromedial gigantocellular zone (RGc) in the reticular formation. On the basis of tracing experiments, the RPc can be further subdivided into a dorsolateral and a ventromedial zone (RPcdl and RPcvm, respectively; fig. 2, 3A); the latter may correspond to the mammalian intermediate reticular formation, the former to the mammalian RPc proper. Caudally, the large cells of RGc disappear and RGc and RPcvm blend into the central nucleus of the medulla oblongata, pars ventralis (CMOv); pars dorsalis (CMOd) is the continuation of RPcdl.

Premotor cells of both the motor nuclei of jaw opening and jaw closing muscles were found throughout RPcvm. In contrast, within RPcdl, premotor cells of jaw opening muscles have been found to be restricted to its caudal compartment, those of jaw closing muscle to its rostral compartment. The intermediate compartment of RPcdl is free of premotor cells of jaw muscles [Bout and Dubbeldam, 1994]. A further difference between RPcdl and RPcvm is that the former has ipsilateral projections to the motor nuclei, whereas the latter has bilateral projections. In mammals, the area intertrigeminalis and lateral tegmental field are sources of projections to trigeminal motor neurons [summarized in e.g. Landgren et al., 1986]. An area intertrigeminalis has been described in birds [Arends and Dubbeldam, 1982], but is considered part of the RPcdl of the present study. As for the motoneurons innervating the facial musculature in mammals, these cells appear to receive afferents from the parvocellular reticular formation [Mogoseanu et al., 1994]. As said before, there is no extensive facial musculature in birds.

A comparable approach can be used for the tongue muscles (fig. 3B): there are three motor pools innervating the extrinsic tongue retractors ($mVII_i$) and tongue protractors (nucleus retrofacialis: mIX_{rf}) and the intrinsic tongue muscles [nucleus nervi hypoglossi, pars lingualis: mXII, 1; Dubbeldam and Bout, 1990]. The intermandibular muscle innervated by the trigeminal motor group mV_3 contributes to dorsoventral movements of the tongue. Again, the situation in birds seems more complicated than that in mammals, as in the latter mainly the hypoglossal nucleus seems to innervate the intrinsic and extrinsic muscles [e.g. Travers et al., 1995]. However, a few reports suggest that in mammals, too, the lingual musculature has a dual innervation by facial and hypoglossal motoneurons [Chibuzo and Cummings, 1983; Satomi et al., 1985]. In birds, the premotor cells of the tongue motor nuclei are found in the intermediate and caudal compartments of RPcdl and throughout RPcvm (fig. 3B) [Tellegen, 1997]. So, there is at least partial overlap of the premotor areas of tongue and jaw musculature. This overlap suggests that reticular premotor neurons in these regions have a coordinating role for jaw and tongue movements.

In this context it is interesting to note that van Gennip [1988] found in the pigeon that tongue retraction is the most stereotyped movement of the tongue during pecking and the various mechanisms of food transport through the oral cavity [throw-and-catch, slide-and-glue; see Zweers, 1982b]. Tongue retraction is coupled to jaw opening, which may be reflected in the cooccurrence of premotor neurons of the two systems in the caudal compartment of RPcdl. In a comparison of three breeds of pigeons with different lengths of the beak, protraction appeared to be more vari-

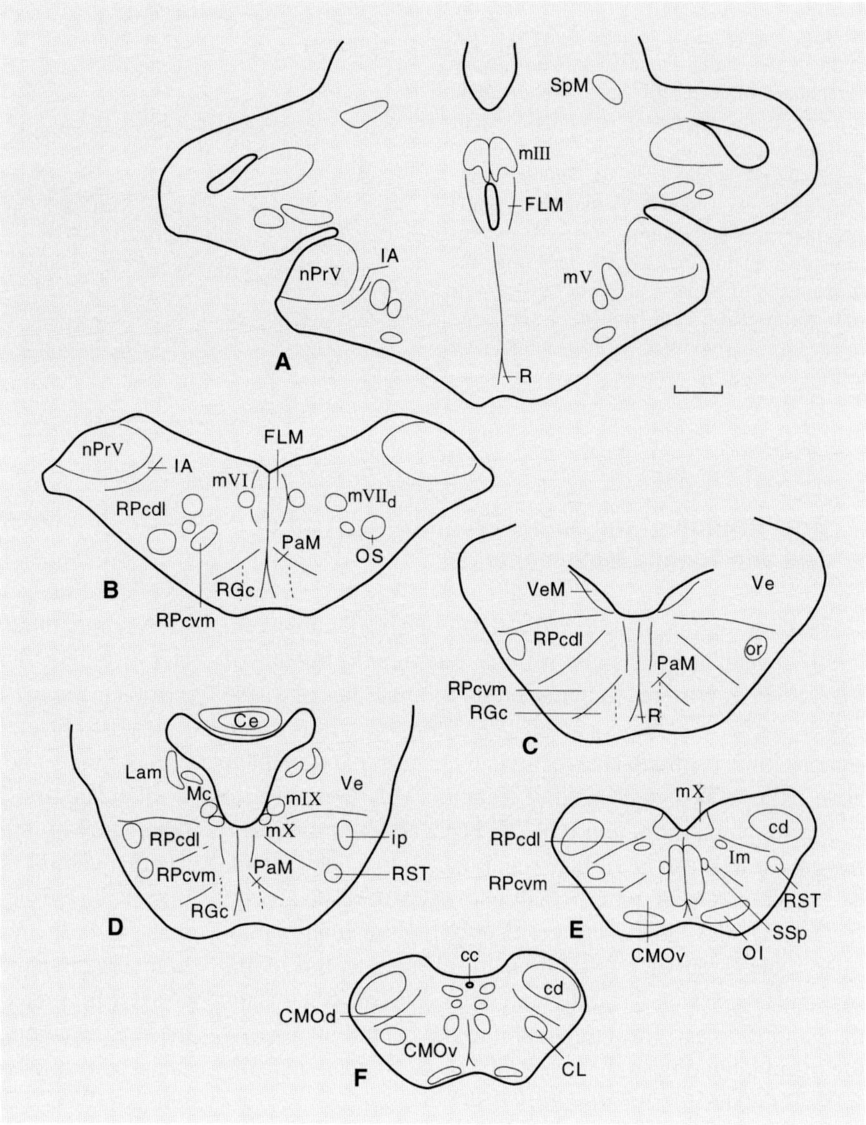

Fig. 2. Schematic cross sections through brainstem of the mallard indicating the borderlines between the longitudinal zones of the reticular formation. **A–F** From rostral to caudal [from Tellegen and Dubbeldam, 1999]. cd = Subn. caudalis of TTD; CL = n. cervicalis lateralis; CMOd, CMOv = n. centralis medullae oblongatae, pars dorsalis, ventralis; FLM = fasciculus longitudinalis medialis; IA = area intertrigeminalis; Im = n. intermedius (dorsal motor nucleus of nervus, XII); ip = subn. interpolaris of TTD; Lam = n. laminaris; Mc = n. cochlearis magnocellularis; mIII, mIV, mV = motornuclei nervus III, IV, V; nPrV = nucleus sensorius principalis trigeminalis; OI = oliva inferior; or = subn. oralis of TTD; OS = oliva superior; PaM = n. paramedialis (part of RGc); R = nucleus raphe; RGc = n. reticularis gigantocellularis; RPcdl, RPcvm = n. reticularis parvocellularis, zona dorsolateralis, ventrolateralis; RST = n. reticularis subtrigeminalis; SpM = n. spiriformis medialis; SSp = n. supraspinalis (ventral motor nucleus of nervus XII); TTD = nucleus and tractus descendens of nervus trigeminus; Ve = area vestibularis; VeM = n. vestibularis medialis.

able and may depend more on mechanoreceptive information [van Gennip, 1988]. In mammals, tongue protrusion and tongue retraction slightly precede jaw opening and jaw closing [Kaku, 1984]; motoneurons for both tongue movements are found in the hypoglossal nucleus. Tomioka et al. [1997] found that the jaw closing reflex is accompanied by protraction of the tongue, both movements being elicited by mechanosensory stimulation. The analysis of straining in ducks reveals that tongue retraction is coupled to beak opening, but may continue into the early beak closing, whereas later beak closing is accompanied with tongue protraction [Zweers et al., 1977]. These observations all demonstrate that tongue and jaw movements are closely coordinated. Such coordination requires a common coordinating system; the reticular formation, more in particular the RPcvm, may be the neural substrate for this coordination.

A Swallowing Center in Birds?

Nothing is known about a swallowing center in birds, though the existence of such a center can be expected. In mammals, swallowing in its strict sense and peristalsis can be distinguished [Jean, 1984]. Swallowing depends upon the coordinated action of muscles of the mouth, pharynx

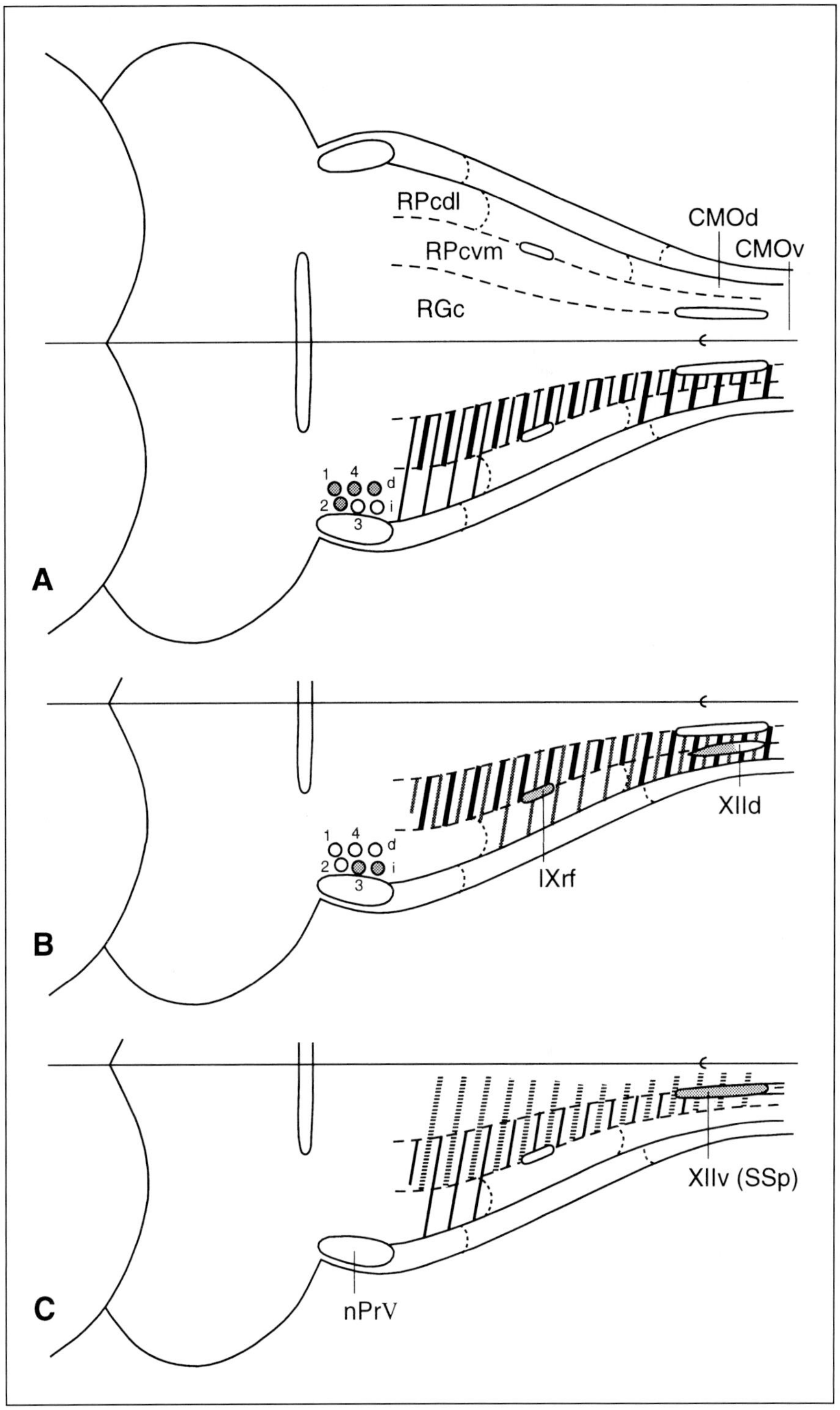

Fig. 3. Diagram of horizontal view showing the distribution of premotor neurons in the reticular formation of the lower brainstem. **A** Hatched areas contain premotor neurons of the jaw opening (thick lines) and jaw closing muscles (thin lines). Also indicated are the cell groups containing motoneurons innervating m. depressor mand. ($mVII_d$: d), m. protractor quadrati (mV_4: 4), adductor muscles (mV_2: 2) and pterygoid muscles (mV_1: 1). **B** Hatched areas contain premotor neurons of jaw opening (thick lines) and tongue muscles (thin lines). Also indicated are the motor nuclei innervating tongue muscles: extrinsic protractors ($mVII_i$: i) and retractors (IXrf), intrinsic muscles (XIId) and intermandibular muscle (mV_3: 3). **C** Hatched areas contain premotor neurons of jaw closing muscles (thin lines) and craniocervical muscles (broken lines) [based on Bout and Dubbeldam, 1994; Tellegen and Dubbeldam, 1994, 1999]. For abbreviations, see fig. 2 and text.

and larynx and is under the control of a medullary swallowing center consisting of two cell groups [Amri et al., 1984; Kessler and Jean, 1985]. One cell group (group 1 neurons) lies dorsal and is a ventral part of the solitary complex, the other cell group (group 2 neurons) is part of the ventromedial medullary reticular formation close the nucleus ambiguus. The latter center may be recipient of projections of the dorsal group and has projections both to the hypoglossal nucleus [Amri and Car, 1988] and the trigeminal motor nucleus [Jean et al., 1983]. Probably the nucleus ambiguus is also part of the swallowing system, as it innervates the laryngeal muscles [e.g. Okubo et al., 1987]. It has been suggested that swallowing depends on a central network, but with the possibility of sensory feedback through the solitary complex [Jean, 1984].

In birds, the oral transport is mainly a function of jaw and tongue muscles. Swallowing involves movements of the larynx and ventral pharyngeal valves [Zweers, 1982b]. Further, Zweers et al. [1981] described the mechanism of the glottis in the pigeon that is important in preventing food from entering the trachea. The laryngeal muscles are innervated by the nervus ambiguus. Little is known about the reticular premotor system in control of the activity of this motor nucleus. However, in birds, too, the solitary complex has an important role in providing sensory feedback. Arends et al. [1988] showed that efferents of the solitary complex enter – judging from their description and pictures – the part of RPc corresponding mainly to our RPcdl, also touching the nucleus ambiguus. This suggests that specifically RPcdl (possibly including CMOd) contains the equivalent of the group 2 neurons of the swallowing network.

Motor Control System for Pecking and Head Movements

Pecking and other feeding movements comprise not only the movements of beak and tongue, but also head and neck movements. So, there should be a center in the control of beak, tongue and head movements.

The muscles responsible for head and neck movements are innervated by motoneurons in the ventral horn of the cervical cord and its rostral continuation in the ventral hypoglossal or supraspinal nucleus [Wild and Zeigler, 1980; Hörster et al., 1990; Zijlstra and Dubbeldam, 1994]. Tracing studies in the mallard revealed that reticular premotor neurons of the supraspinal nucleus occur in RPcvm and in the gigantocellular reticular formation of the brainstem (RGc, fig. 3C) [Tellegen and Dubbeldam, 1999]. Double-labeling studies showed that labeled cells occur in the same parts of RPcvm after tracer injections in the jaw motor nuclei and supraspinal nucleus, respectively, but hardly any double-labeled cells were found [Tellegen and Dubbeldam, 1996]. The occurrence of a few double-labeled cells may be caused by the fact that it is very hard to confine the injections to the motor nuclei, so that the injection may cover a small part of the surrounding reticular formation. Anyhow, these observations strongly suggest that RPcvm takes part in the coordination of beak and head movements, whereas RPcdl and RGc may be involved with more specific aspects of the respective mechanisms within the general pattern of movements. This could mean that RPcvm is the seat of the movement pattern-generating system.

Unfortunately, little is known about the physiology of this region in birds, but comparison with the mammalian situation may help to elucidate the situation.

A Masticatory Rhythm Generator

Pecking in birds may seem a less rhythmic movement than the process of biting and chewing in mammals. However, the sequence of pecking and mandibular transport requires a series of rather similar jaw and tongue movements [for more details, see accompanying paper by Bout, 1998]. Moreover, some feeding types such as, for example, straining in ducks consist of repetitions of fast uniform jaw movements. So far, nothing is known about the existence or organization of a rhythm generator in birds. It may help, however, to see what data are available about such a rhythm-generating system and its neuronal substrate in mammals.

Over the years Nakamura and Katakura [1995] developed a model for a rhythm generator for masticatory movements. Even though the details of the model may need further clarification, it is interesting to focus on two of its aspects. The first is that parts of the gigantocellular and paragigantocellular reticular formation appear to be involved, with the parvocellular reticular formation – equivalent of the avian RPc – serving as a source of input to the trigeminal motor nuclei (fig. 4). Apparently, different cells project to the jaw opening and jaw closing motor nuclei, but a compartmentalization as found in the avian RPcdl has not been mentioned. In birds, little is known about connections between the respective parts of the reticular formation; there is evidence of reciprocal projections between RGc and RPcvm and between RPcvm and RPcdl, but not between RPcdl and RGc [Bout and Dubbeldam, 1994]. The

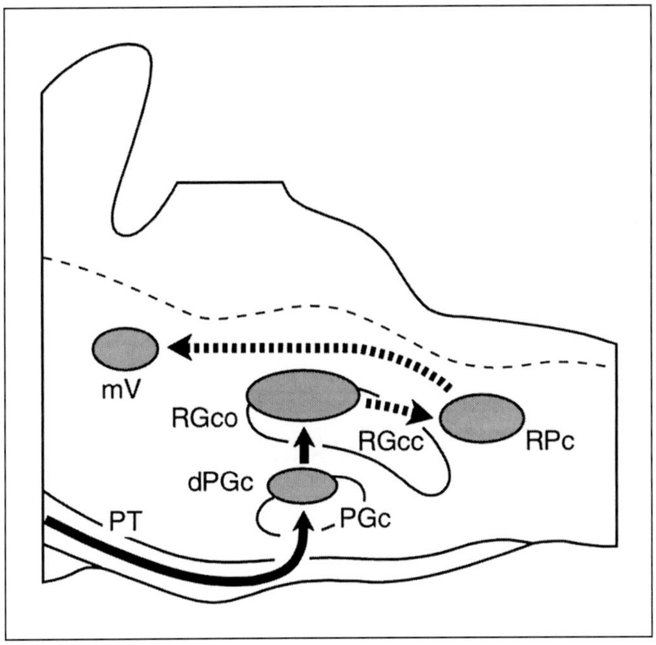

Fig. 4. Schematic representation of the neural substrate of the mammalian rhythm generator for chewing. Redrawn after Nakamura and Katakura [1995], slightly modified. dPGc = Pars dorsalis of PGc; PGc = n. reticularis paragigantocellularis; PT = pyramidal tract; RGcc, RGco = pars caudalis, pars rostralis of RGc (further see fig. 2).

second aspect is that in mammals the rhythm-generating system is under telencephalic control through the corticobulbar input [fig. 4) [e.g. Chandler and Tal, 1986]. In birds, part of the archistriatum is the source of a large extratelencephalic pathway, the occipitomesencephalic tract. This archistriatal part is considered a sensorimotor center. The occipitomesencephalic tract has intense terminations in RPcdl, more moderate terminations in RPcvm, but the RGc is devoid of such a projection (fig. 5) [Dubbeldam et al., 1997].

On account of the assumed coordinating role of RPcvm, suggested in the preceding sections, this area can be assumed to represent the avian equivalent of the mammalian parvocellular part of a pattern-generating system. The most important difference from the mammalian model is that RPcvm receives a direct, ipsilateral telencephalic projection and not an indirect one through RGc. In this context it can be noted that, according to Landgren et al. [1986], the mammalian intertrigeminal area is a second recipient of cortical input and source of a projection to the trigeminal motor nucleus; the avian intertrigeminal area is part of RPcdl (see above). Thus there may be distinct parallels between the two groups of vertebrates, but recognition of the parallels are hampered by the difficulty to homologize more precisely the subdivisions of the reticular formation.

Individual animals may have stereotyped patterns of mastication, but differences in performance do occur in connection with differences in texture and size of food particles [Lund and Enomoto, 1988]. In a study on the rat van Willigen et al. [1986] suggested that oral activity may be controlled by two relatively independent systems, a program-oriented and an environment-oriented system. The first system seems to correspond with the pattern-generating system, whereas the second one should at least be recipient of exteroceptive and proprioceptive information. Then, the question can be raised as to whether it is possible to distinguish the two systems within the organization of the neural premotor substrate using sensory input as a criterion.

Sensory Input to the Reticular Premotor Areas

Proprioceptive System of the Jaw Muscles

The mesencephalic nucleus of the trigeminal system (mesV) in birds innervates the muscle spindles of the jaw closing muscles. The jaw opening and extrinsic tongue muscles do not possess muscle spindles [Bout and Dubbeldam, 1991]; nothing is known about the occurrence of muscle spindles in the intrinsic tongue muscles. mesV has direct projections to the motor nuclei of the jaw closing muscles and further sends – at least partly collateral – projections to the rostral and caudal compartments of the RPcdl (fig. 6) [Bout et al., 1997]. Lucchi et al. [1997] found that more caudal projections from mesV derive from a separate population of cells. Anyhow, premotor areas of both the jaw opening and jaw closing muscles appear to receive proprioceptive input from the jaw closing muscles.

Even though the overall picture is the same for birds and mammals [e.g. Rokx et al., 1986], there are two differences. The first is that birds lack teeth and do not have periodontal sense organs innervated by mesV. In the second place, even though it has been mentioned in the literature [e.g. Bortolami et al., 1972], there is no definite evidence of a cerebellar projection from mesV. This could mean that the mesV projection to the cerebellum in mammals is connected with the presence of the second proprioceptive component, the periodontal sense organs. In birds, the muscle spindles of the jaw muscles may have mainly a role in the control of the gape of the beak [Dubbeldam, 1985; Bout and Dubbeldam, 1991].

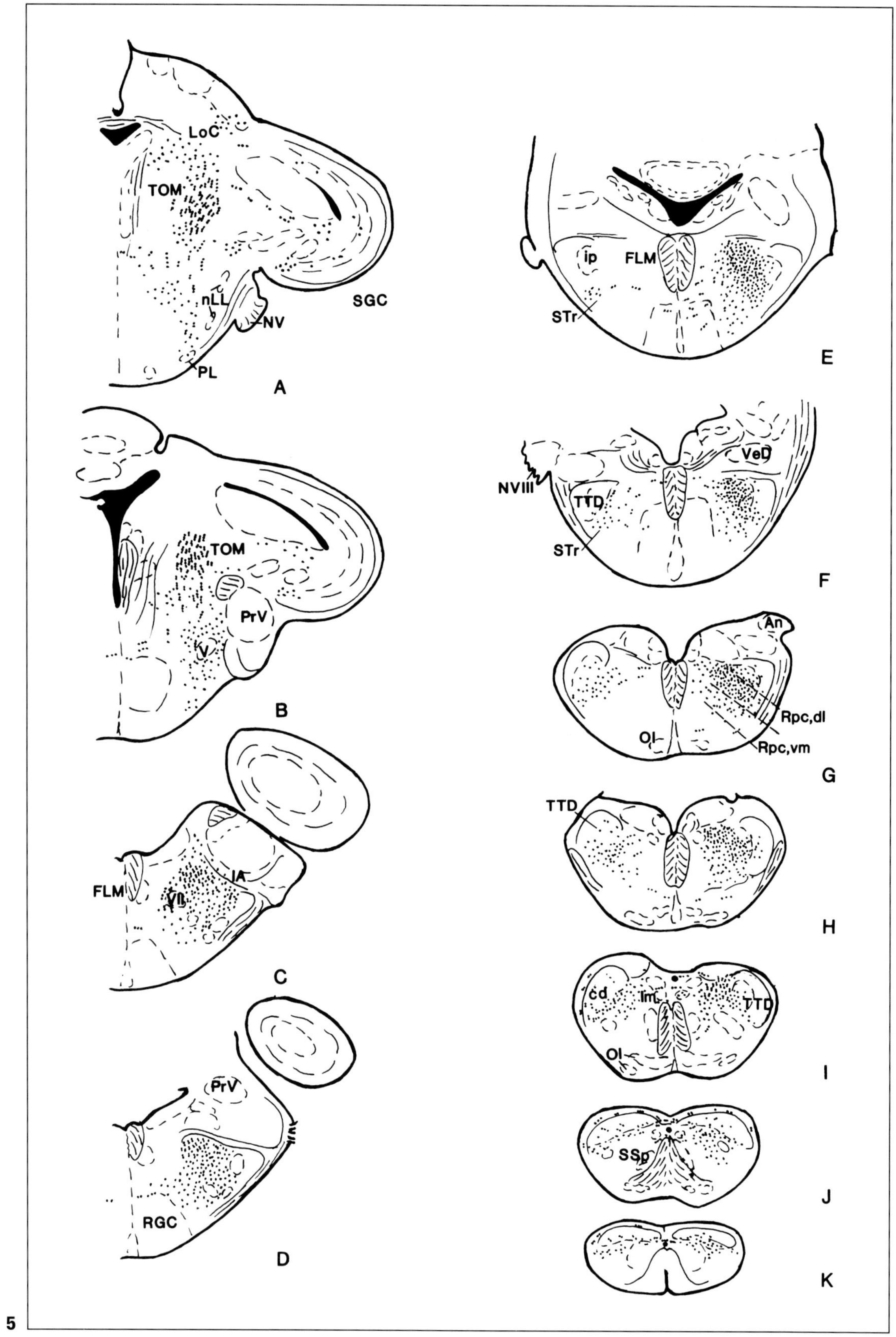
LoC
TOM
nLL
NV
SGC
PL
A
TOM
PrV
V
B
FLM
VII
IA
C
PrV
RGC
D
ip
FLM
STr
E
VeD
NVIII
TTD
STr
F
An
Rpc,dl
Ol
Rpc,vm
G
TTD
H
cd
Im
TTD
Ol
I
SSp
J
K
5

There is little information about the presence of other types of proprioceptive sense organs in the jaw and tongue systems of birds, but electrophysiological recordings in the pigeon indicate that the descending trigeminal system does receive proprioceptive input [Silver and Witkovsky, 1973]. However, it is not possible to distinguish specific proprioceptive cell groups within the sensory trigeminal system. We will consider this system in more detail in the next section.

Exteroceptive Input to the Bulbar Reticular Formation

Several sensory systems send input to parts of the bulbar reticular formation. Data about three sensory systems is summarized in figure 7.

Trigeminal Sensory System. Tactile sense is important in the detection and recognition of food, but also for the control of transport of food through the oral cavity [e.g. Berkhoudt, 1980; Zweers, 1982b]. Tactile information from the beak is conveyed by the trigeminal nerve [e.g. Dubbeldam and Karten, 1978], that from the tongue through the glossopharyngeal [Dubbeldam et al., 1979] and hypoglossal nerve [Wild, 1990]. These nerves project to the sensory trigeminal nuclei, i.e. the principal sensory nucleus (projects to the forebrain) and the subnuclei of the descending trigeminal system. The subnucleus oralis projects to the cerebellum; RPcdl and CMOd receive projections from the subnuclei interpolaris and caudalis (fig. 7) [Arends and Dubbeldam, 1984]. These subnuclei may convey both proprioceptive and exteroceptive information to the premotor systems of jaw and tongue muscles.

Visual System. The tectum mesencephali is an important recipient of retinal input and source of (among others) two descending pathways: an ipsilateral pathway to the lateral pontine nucleus and surrounding reticular formation and a decussating pathway with projections to the medial RGc [nucleus paramedianus; Reiner and Karten, 1982: pigeon; Masino and Knudsen, 1992: barn owl; Tellengen et al., 1998: mallard]. The latter projection suggests that visual information is used for the control of head and neck movements and other activities under the control of RGc (see

Fig. 5. Telencephalic projections to parts of the reticular formation/reticular premotor systems [after Dubbeldam et al., 1997]. LoC = Locus ceruleus; nLL = n. lemnisci lateralis; NV = nervus trigeminus; PL = n. pontis lateralis; SGC = stratum griseum centrale; STr = n. reticularis subtrigeminialis; TOM = tractus occipitomesencephalicus (further see fig. 2).

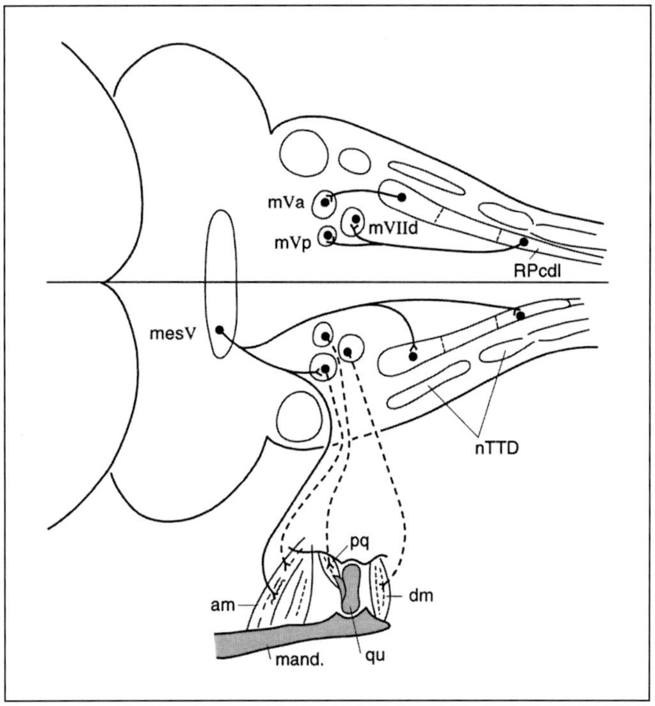

6

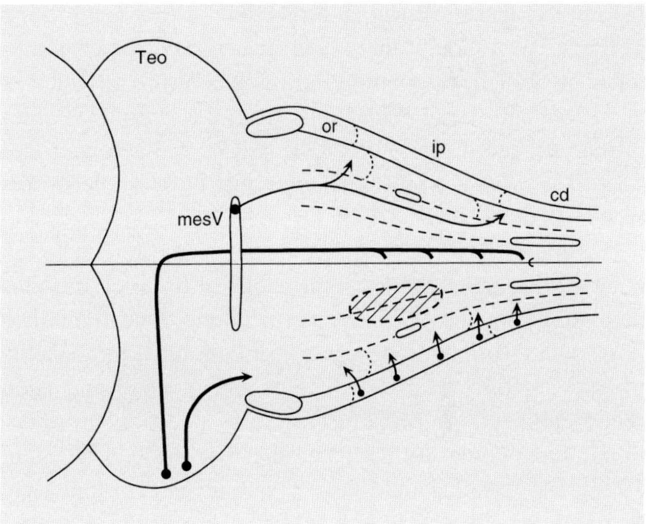

7

Fig. 6. Proprioceptive system of jaw muscles. Neurons of the mesencephalic trigeminal nucleus (mesV) innervate the muscle spindles in jaw closing muscles and have projections to the motor nuclei (mVa) innervating these muscles and to the rostral and caudal compartment of RPcdl. Jaw opening muscles do not have muscle spindles and their motor nuclei (mVp, mVIId) do not get an input from mesV. Broken lines represent motor nerves (for other abbreviations, see legends of fig. 1, 2).

Fig. 7. Proprioceptive and exteroceptive input to the reticular premotor systems: mechanoreceptive input (arrows) from two nuclei of the descending trigeminal system (ip, cd) and from mesV (thin lines), vestibular input (hatched area) and ipsilateral and contralateral input from the optic tectum (Teo, thick lines). Compare figure 3 for subdivisions of the reticular formation (for abbreviations, see fig. 2).

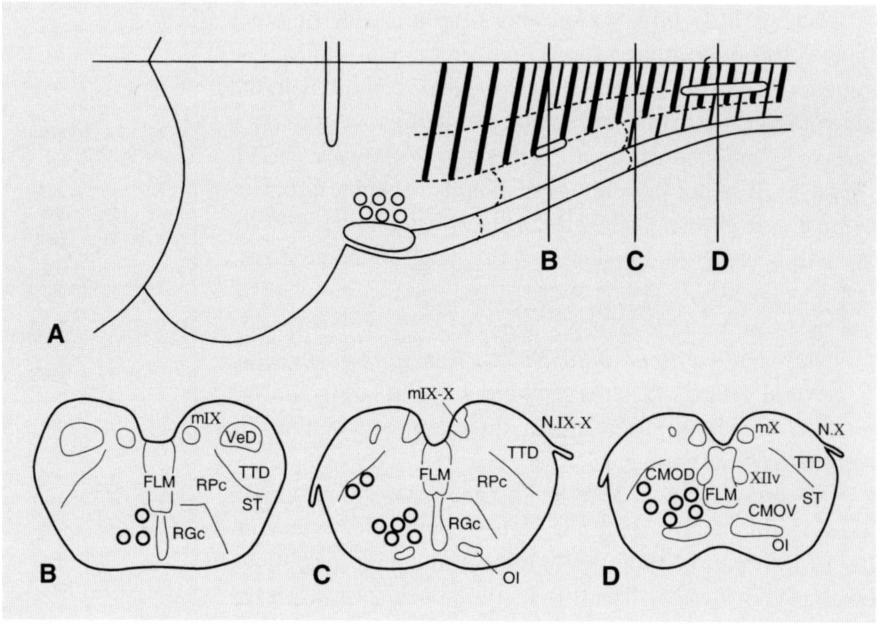

Fig. 8. A Horizontal diagram of reticular formation showing the premotor regions for head and neck movements (thick lines) and locomotion (thin lines). **B–D** Cross sections through brainstem of an anseriform bird showing locations where locomotion can be evoked by electrophysiological stimulation [modified after Steeves et al., 1987]. For abbreviations, see fig. 2.

below). So far, a tectospinal projection has not been found in birds. Even though there is clear evidence for visuomotor control of beak opening during pecking [e.g. Deich et al., 1985], no direct projections from the tectum or other primary visual centers to RPcdl or RPcvm have been described. Possibly, telencephalic circuits have a role in this type of control [Jäger, 1990].

Vestibular System. Vestibular nuclei send efferents to the oculomotor nuclei and to motor nuclei of the neck muscles [e.g. Arends et al., 1991]. Such projections serve to mediate positional information necessary for corrective movements of eyes (stabilization visual field) and of head and neck (head stabilization). More recently, also a vestibular projection to the reticular formation has been described [Tellegen, 1997]; the terminal field covers part of RPcvm and RGc that also contains premotor neurons for jaw and neck movements (fig. 7). This suggests that vestibular information can be used in feeding behavior; this type of information may be more important for the system in control and head and neck movements than that in control of jaw and tongue movements. In addition to the above-mentioned projections a vestibulospinal tract exists. This one, too, may have a role in the control of head and neck movements, but also in the control of body posture and locomotion.

Bulbar Premotor System for Locomotion and Head Movements

Walking of flying birds need to stabilize the position of the head. In other words, a coordination of locomotory activity and head movements is required. Using electrostimulation Steeves et al. [1987] charted the location in the pontomedullary reticular formation of anseriform birds, where locomotion can be evoked (fig. 8). Areas corresponding to our CMOv, extending into the caudal half of RGc, and CMOd appear to be parts of the locomotor pathways. CMOv does not receive projections from the occipitomesencephalic tract [Dubbeldam et al., 1997], but the basal branch of the septomesencephalic tract originating in the rostral Wulst appears to send a few fibers into this area [Dubbeldam, 1976].

Premotor neurons of the locomotor system and of the neck system coincide in CMOv and RGc (fig. 8A). It is specifically RGc that is the recipient of the decussating tectal input. So, the visual information conveyed by the tectobulbar system may, in particular, be important for locomotion and connected activities.

Table 1. Importance of the control of position and/or movement of a few morphological elements during different activities

Control of	Feeding	Vocalization	Respiration
Gape beak	+	+	±
Tongue	+	+	–
Larynx	+	±	+
Head position	+	+	–

+ = Important; ± = less important; – = unimportant.

Table 2. Overview of several muscle systems, their motor nuclei and related reticular premotor systems

Muscle group	Motor nuclei	Premotor area
Jaw muscles[a]		
Jaw opening muscles	mV_4, $mVII_d$	RPcvm/$RPcdl_c$
Jaw closing muscles	$mV_{1,2}$	RPcvm/$RPcdl_r$
Tongue muscles		
Extrinsic		
'Retractors'	$mVII_i$	RPcvm/$RPcdl_{i,c}$
'Protractor'	mIX_{rf}	RPcvm/$RPcdl_{i,c}$
Intermandibular	mV_3	RPcvm/$RPcdl_{i,c}$
Intrinsic	$mXII_d$ (XII, l)	RPcvm/$RPcdl_{i,c}$
Laryngeal muscles	mX_v (ambiguus)	RPcdl/CMOd?[a]
Tracheal/syringeal muscles	$mXII_d$ (XII, ts)	rVRG/RAm
Neck muscles	$mXII_v$ (SSp) ventral horn	RPcvm/RGc

$mV_{1,2,etc.}$ = Trigeminal motor nucleus, subnucleus 1, 2, etc; $mVII_{d,i}$ = facial motor nucleus, dorsal, intermediate subnucleus; mIX_{rf} = nucleus retrofacialis; $mXII_{d,v}$ = hypoglossal motor nucleus, dorsal, ventral subnucleus; XII, l, ts = dorsal hypoglossal motor nucleus, pars lingualis, tracheosyringealis; RAm = nucleus retroambigualis; rVRG = rostral ventral respiratory group [Reinke and Wild, 1997]; $RPcdl_{r,i,c}$ = rostral, intermediate and caudal compartment of RPcdl.

[a] No definite evidence.

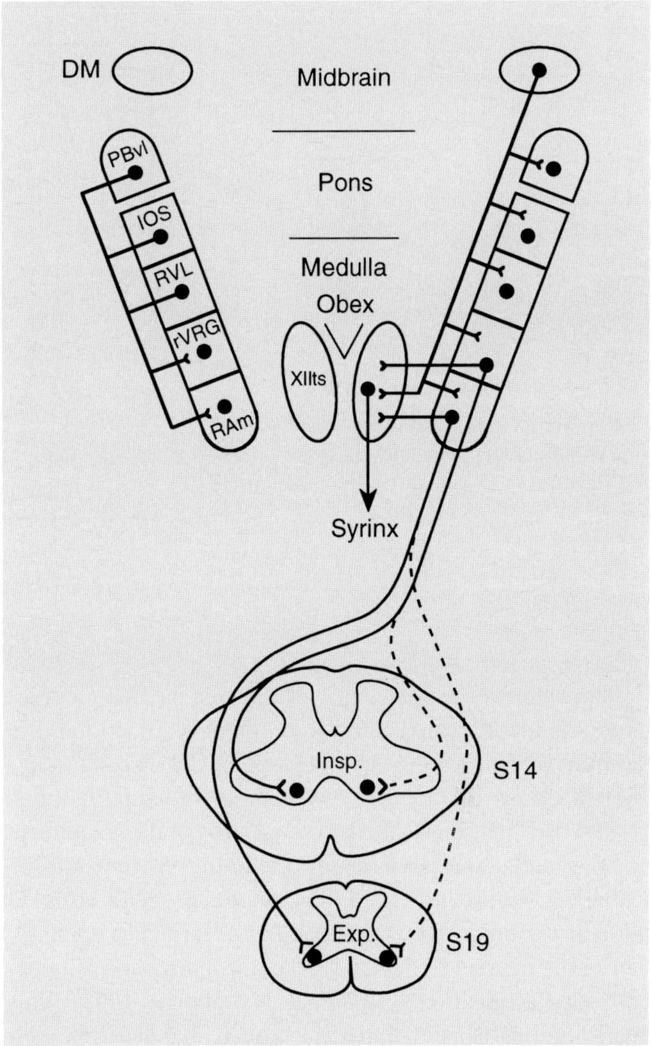

Fig. 9. Summary showing the pontomedullary centers involved in the control of respiration and vocalization [modified after Reinke and Wild, 1997]. DM = Dorsomedial nucleus of intercollicular complex; Insp, Exp = spinal inspiratory, expiratory centers; IOS = n. infraolivaris superior; PBvl = n. parabrachialis ventrolateralis; RAm = n. retroambigualis; RVL = ventrolateral nucleus of rostral medulla; rVRG = rostral, ventral respiratory group; XIIts = nervi hypoglossi, pars tracheosyringealis.

Are Premotor Systems for 'Learned' and 'Nonlearned' Activities Different?

The survey of the preceding sections shows that it is possible to make a functional subdivision of the pontobulbar reticular formation using the pattern of afferent and efferent connections in addition to cytomorphological characteristics as criteria. Reticular premotor areas appear to be the main sources of input to the motor nuclei of several systems in the head. These reticular areas may contain the pattern generators responsible for the basic patterns of movement of the various mechanisms. At the same time, these premotor systems allow a high degree of flexibility in the performance of these mechanisms [Bout, 1998]. Do these premotor systems also allow the emergence of new motor skills? In an attempt to answer this question, we will consider the vocalization system of birds.

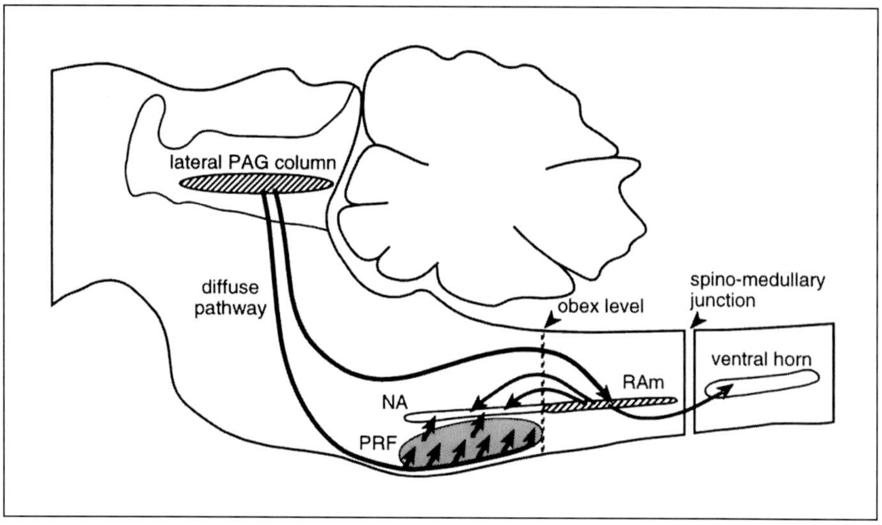

Fig. 10. Diagram showing the pontomedullary centers in control of respiration and vocalization in a mammal [from Zhang et al., 1995, slightly modified].

The interesting aspect of vocalization is that in most birds it is a nonlearned activity, i.e. there is some kind of a training process during the development of vocalizations, but perception of the bird's own song is sufficient. In a few groups of birds, most notably song birds and psittaciform birds, vocalization develops in a learning process, during which the young birds need to be tutored by adult birds. In these birds song can be considered as a learned motor skill; the learning process depends upon the presence of a specific song acquisition system [e.g. Nottebohm, 1993]. This vocal learning is an evolutionary novelty that developed independently in several groups of birds.

Two points are important for the following discussion of the brainstem structures involved in vocalization. In the first place, the mechanism of sound production requires a precise regulation of the respiration and a good control of the activity of the muscles of the sound-producing organ, the syrinx. So, there should be a strict coordination of the activity of the respiratory and syringeal muscle centers, even though a certain degree of motor noise may exist [Suthers et al., 1996]. We will have a closer look at the neural substrate for respiration in the next section.

In the second place, a singing bird may have to sustain a certain gape of the beak during vocalization, may keep its head in a specific position, may use its tongue, regulate glottal opening and so on [e.g. Westneat et al., 1993]. In other words, during vocalizations the birds use the same elements as during feeding or drinking behavior and preening (table 1). Then, it can be expected that the vocalization system and the feeding system share elements of the motor and premotor centers in the brainstem (table 2). An important difference between the motor nuclei of the syringeal and laryngeal muscles and the other motor nuclei of the brainstem is that in songbirds and budgerigars the first two receive a direct telencephalic input from an archistriatal nucleus: the nucleus robustus in song birds [e.g. Nottebohm, 1993; Wild, 1993b] and nucleus centralis archistriatalis in the budgerigar [Striedter, 1994]. Such a nucleus has not been found in other groups of birds. This means that this acquired motor activity is under direct control of a telencephalic programming network [e.g. Vu et al., 1994].

The Coherence of Respiratory and Vocalization System

Recently, Reinke and Wild [1997] described the respiratory pathway of the pigeon in detail. Figure 9 summarizes some of their observations: important reticular centers in the brainstem are – from caudal to rostral – the nucleus retroambigualis (RAm) and the rostral ventral respiratory group (rVRG) with projections to spinal expiratory and inspiratory centers, respectively; the ventrolateral nucleus of the rostral medulla, nucleus infraolivaris superior and nucleus parabrachialis ventrolateralis, all sources of projections to RAm and rVRG. These nuclei all receive input from the intercollicular dorsomedial nucleus that also sends a projection to the motor nucleus of the syringeal muscles (mXIIts). Repeatedly, the intercollicular center has been considered the highest center for evoking nonlearned vocalizations [calls; discussion in Wild, 1997]. Recently, Goller and Larsen [1997] propose a new mechanism for the syringeal function. In this context these authors suggest that the functioning of the syringeal mechanism and respiration are decoupled to a certain degree. This may be reflected by the fact that the archistriatal nucleus robustus comprises subdivisions that control the intercollicular respiratory center and mXIIts, respectively [Vicario, 1991]. Then, also intercollicular projections to the respiratory centers and the syringeal motor center may derive from different cell populations.

Comparing the avian pattern of respiratory pathways with that in mammals, it is interesting to note that basically a similar pattern of cell groups and projections seems to exist (fig. 10). An important feature is that the retroambigual nucleus receives input from the periaquaductal gray [e.g. Gerrits and Holstege, 1996]. It is tempting to assume that the dorsomedial nucleus of the intercollicular complex, one of the sources of input to the nucleus retroambigualis in birds [Vicario, 1993; Wild, 1993a], is the functional equivalent of those lateral parts of the mammalian periaquaductal gray that project upon nucleus retroambigualis. Zhang et al. [1995] ascribe the ventral reticular formation and retroambigual nucleus a role in the control of vocalization through projections to the nucleus ambiguus that innervates the laryngeal muscles [Okubo et al., 1987]. In this context it is interesting to note that a discrete zone of the periaquaductal gray has also a direct projection to nucleus ambiguus and thus may have a role in vocalization [Ennis et al., 1997].

Mammals do not possess a syringeal system, whereas in birds the role of the larynx may be less important [but not unimportant, see Brauth et al., 1997]. In birds, the laryngeal musculature is innervated by the nucleus ambiguus [Wild, 1981]; so far, there is no evidence that this center receives a direct input from the dorsomedial cell group of the intercollicular nucleus. A further difference between birds and mammals, of course, concerns the ventilation mechanism of the lungs. However, even though certain differences exist between the two groups, the respiratory systems of birds and mammals appear to have several traits in common [also see discussion in Wild, 1997].

Concluding Considerations

The preceding overview leads to a few considerations and suggestions.

(1) A specific musculomotor system is used for several different activities, e.g. jaw muscles are used in feeding, preening and vocalization. The motor activity for a specific activity will often have a basic pattern, but this pattern is modifiable on the base of available sensory information. The basic patterns need not be the same for different activities. Should a premotor system contain more than one (type of) pattern generator, as it may serve for diverging motor functions? The answer may be that changing configurations of neurons within a premotor system participate in generating the varying patterns of motor activity. This precludes the existence of rigidly working rhythm generators.

(2) A specific activity can have different forms, e.g. different types of feeding in birds exist: pecking, straining, probing or seed husking. Assuming basically stereotyped motor patterns generated by a premotor system, can this motor pattern be used by a mechanical system, or by differently built mechanical systems in different birds, to perform different types of activities? On account of consideration 1 a reasonable assumption is that the evolution of movement patterns depends upon changes of both peripheral mechanisms and motor patterns.

(3) Different activities require the activity of differing combinations of muscle groups and thus of different combinations of motor nuclei and related premotor systems, e.g. jaw, tongue and neck muscles in feeding, and jaw, tongue, laryngeal and syringeal muscles plus respiratory muscles in vocalization. In other words, motor patterns can be used in different combinations and changes within these combinations – e.g. of timing of the activity of different premotor systems – may result in the emergence of new activities. An example of such a shift in timing may be the existence of tip-up drinking in Bengalese finches versus tip-down drinking in Zebra finches [Heidweiller and Zweers, 1990].

(4) According to Baev [1997], ‘learning is a process that leads to the appearance of new automatisms or to the improvement or refinement of one that already exists’. In his view ‘plasticity and learning are principally similar processes’, albeit with different temporal duration. The concept of the modal action pattern, mentioned in the introduction of the present paper, may include both aspects: modifiability of motor patterns would allow new patterns or combinations of patterns and open the possibility for learning processes resulting in improved motor performance

(e.g., in terms of pecking efficiency), as well as the emergence of (evolutionary) new patterns. Learning at the level of the pontobulbar reticular formation may imply primarily the refining of already existing motor patterns. In cases where such a basic pattern generator does not exist, learning may depend upon the involvement of hierarchically organized premotor systems. This may be the main difference between 'nonlearned' and 'learned' vocalizations, e.g. calls versus song in songbirds. This example is not unique.

(5) Specific premotor systems have a specific localization, even though the location of different premotor systems can overlap. On the other hand, closely connected activities may be under the control of spatially segregated premotor systems. For example, Balaban [1997] working with chicken-quail chimeras showed that vocalizations (crows) and connected head movements (head bobbing) are under control of different parts of the brainstem. Bass and Baker [1997] suggest that specifically rhythmic patterns generating circuits arise from rhombomeres 7 and 8. Our data suggest that in birds and mammals more rostral parts may also be involved.

Acknowledgement

I thank Dr. Ron Bout and two reviewers for their valuable comments on a previous draft of this paper. Mr. Martin Brittijn prepared the figures used in this study.

References

Amri, M., A. Car (1988) Projections from the medullary swallowing center to the hypoglossal motor nucleus: A neuroanatomical and electrophysiological study in sheep. Brain Res *441:* 119–126.

Amri, M., A. Car, A. Jean (1984) Medullary control of the pontine swallowing neurones in sheep. Exp Brain Res *55:* 105–110.

Arends, J.J.A., R.W. Allen, H.P. Zeigler (1991) Organization of the cerebellum in the pigeon *(Columba livia).* III. Corticovestibular connections with eye and neck premotor areas. J Comp Neurol *306:* 273–289.

Arends, J.J.A., J.L. Dubbeldam (1982) Exteroceptive and proprioceptive afferents of the trigeminal and facial motor nuclei in the mallard (*Anas platyrhynchos* L.). J Comp Neurol *209:* 313–329.

Arends, J.J.A., J.L. Dubbeldam (1984) The subnuclei and primary afferents of the descending trigeminal system in the mallard (*Anas platyrhynchos* L.). Neuroscience *13:* 781–795.

Arends, J.J.A., J.M. Wild, H.P. Zeigler (1988) Projections of the nucleus of the tractus solitarius in the pigeon *(Columba livia).* J Comp Neurol *278:* 405–429.

Baev, K.V. (1997) Highest level automatisms in the nervous system: A theory of functional principles underlying the highest forms of brain function. Prog Neurobiol *51:* 129–166.

Balaban, E. (1997) Changes in multiple brain regions underlie species differences in a complex, congenital behavior. Proc Natl Acad Sci USA *94:* 2001–2006.

Bass, A.H., R. Baker (1997) Phenotypic specification of hindbrain rhombomeres and the origins of rhythmic circuits in vertebrates. Brain Behav Evol *50(suppl):* 3–16.

Berkhoudt, H. (1980) The morphology and distribution of cutaneous mechanoreceptors (Herbst and Grandry corpuscles) in bill and tongue of the mallard (*Anas platyrhynchos* L.). Neth J Zool *30:* 1–34.

Bortolami, R., E. Callegari, M.L. Lucchi (1972) Anatomical relationship between mesencephalic trigeminal nucleus and cerebellum in the duck. Brain Res *47:* 317–329.

Bout, R.G. (1998) Complex movement patterns: Modifiability and constraints. Acta Anat *163:* 144–156.

Bout, R.G., J.L. Dubbeldam (1991) Functional morphological interpretation of the distribution of muscle spindles in the jaw muscles of the mallard *(Anas platyrhynchos).* J Morphol *210:* 215–226.

Bout, R.G., J.L. Dubbeldam (1994) The reticular premotor neurons of the jaw muscle motor nuclei in the mallard (*Anas platyrhynchos* L.). Eur J Morphol *32:* 134–137.

Bout, R.G., A.J. Tellegen, J.L. Dubbeldam (1997) Central connections of the nucleus mesencephalicus nervi trigemini in the mallard (*Anas platyrhynchos* L.). Anat Rec *248:* 551–562.

Bramble, D.M., D.B. Wake (1985) Feeding mechanisms of lower tetrapods; in Hildebrand M, DM Bramble, KF Liem, DB Wake (eds): Functional Vertebrate Morphology. Cambridge, Belknap Press, pp 230–261.

Brauth, S.E., J.T. Heaton, S.D. Shea, S.H. Durand, W.S. Hall (1997) Functional anatomy of forebrain vocal control pathways in the budgerigar *(Melopsittacus undulatus);* in Carter CS, II Lederhendler, B Kirkpatrick (eds): The Integrative Neurobiology of Affiliation. Ann NY Acad Sci *807:* 368–385.

Chandler, S.H., M. Tal (1986) The effects of brain stem transections on the neuronal networks responsible for rhythmical jaw muscle activity in the guinea pig. J Neurosci *6:* 1831–1842.

Chibuzo, G.A., J.F. Cummings (1983) An enzyme tracer study of the organization of the somatic motor center for the innervation of different muscles of the tongue: Evidence of two sources. J Comp Neurol *205:* 273–281.

Deich, J., B. Klein, H.P. Zeigler (1985) Grasping in the pigeon: Mechanisms of motor control. Brain Res *337:* 362–367.

Den Boer, P.J., R.G. Bout, J.L. Dubbeldam (1986) Topographical representation of the jaw muscles within the trigeminal motor nucleus. An HRP study in the mallard, *Anas platyrhynchos*. Acta Morphol Neerl Scand *24:* 1–17.

Dicke, U., G. Roth (1998) Evolution of tongue mechanisms and the neural substrate for motor control of feeding in amphibians. Acta Anat *163:* 127–143.

Dubbeldam, J.L. (1976) The basal branch of the septomesencephalic tract in the mallard, *Anas platyrhynchos* L. Acta Morphol Neerl Scand *14:* 98.

Dubbeldam, J.L. (1985) Brainstem mechanisms for feeding in birds: Interaction or plasticity. A functional-anatomical consideration of the pathways. Brain Behav Evol *25:* 85–98.

Dubbeldam, J.L., R.G. Bout (1990) The identification of the motor nuclei innervating the tongue muscles in the mallard *(Anas platyrhynchos);* an HRP study. Neurosci Lett *119:* 223–227.

Debbeldam, J.L., E.R. Brus, S.B.J. Menken, S. Zeilstra (1979) The central projections of the glossopharyngeal and vagus ganglia in the mallard, *Anas platyrhynchos* L. J Comp Neurol *183:* 149–168.

Dubbeldam, J.L., A.M. Den Boer-Visser, R.G. Bout (1997) Organization and efferent connections of the archistriatum of the mallard, *Anas platyrhynchos* L.: An anterograde and retrograde tracing study. J Comp Neurol *388:* 632–657.

Dubbeldam, J.L., H.J. Karten (1978) The trigeminal system in the pigeon *(Columba livia)*. I. Projections of the gasserian ganglion. J Comp Neurol *180:* 661–678.
Ennis, M., S.-J. Xu, T.A. Rizvi (1997) Discrete subregions of the rat midbrain periaquaductal gray project to nucleus ambiguus and the periambigual region. Neuroscience *80:* 829–845.
Gerrits, P.O., G. Holstege (1996) Pontine and medullary projections to the nucleus retroambiguus: A wheat germ agglutinin-horseradish peroxidase and autoradiographic tracing study in the cat. J Comp Neurol *373:* 173–185.
Goller, F., O.N. Larsen (1997) A new mechanism of sound generation in birds. Proc Natl Acad Sci USA *94:* 14787–14791.
Heidweiller, J., G.A. Zweers (1990) Drinking mechanisms in the zebra finch and the Bengalese Finch. Condor *92:* 1–28.
Hörster, W., A. Franchini, S. Daniel (1990) Organization of neck muscle motorneurons in the cervical spinal cord of the pigeon. Neuroreport *1:* 93–96.
Jäger, R. (1990) Visuomotor feeding perturbations after lateral telencephalic lesions in pigeons. Behav Brain Res *40:* 73–80.
Jean, A. (1984) Control of the central swallowing program by inputs from the peripheral receptors. A review. J Auton Nerv Syst *10:* 225–233.
Jean, A., M. Amri, A. Calas (1983) Connections between the ventral medullary swallowing area and the trigeminal motor nucleus of the sheep studied by tracing techniques. J Auton Nerv Syst *7:* 87–96.
Kaku, T. (1984) Functional differentiation of hypoglossal motoneurons during the amygdaloid or cortically induced rhythmical jaw and tongue movements in the rat. Brain Res Bull *13:* 147–154.
Kessler, J.P., A. Jean (1985) Identification of the medullary swallowing regions in the rat. Exp Brain Res *57:* 256–263.
Landgren, S., K.Å. Olsson, K.G. Westberg (1986) Bulbar neurones with axonal projections to the trigeminal motor nucleus in the cat. Exp Brain Res *65:* 98–111.
Lauder, G.V. (1991) Biomechanics and evolution: Integrating physical and historical biology in the study of complex systems; in Rayner JMV, RJ Wooton (eds): Biomechanics in Evolution. SEB Seminar Series 36. Cambridge, Cambridge University Press, pp 1–20.
Lucchi, M.L., P.A. Scapolo, A.M. Barazzoni, P. Clavenzani, G. Lalatta Costerbosa, P. Berardinelli, R. Bortolami (1997) Mesencephalic trigeminal nucleus neurons supplying the jaw closing muscles have no spinal projection: A fluorescent double-labeling study in birds and mammals. Anat Rec *249:* 255–258.
Lund, J.P., S. Enomoto (1988) The generation of mastication by the mammalian central nervous system; in Cohen AH, S Rossignol, S Grillner (eds): Neural Control of Rhythmic Movements in Vertebrates. New York, Wiley, pp 41–72.
Masino, T., E.I. Knudsen (1992) Anatomical pathways from the optic tectum to the spinal cord subserving orienting movements in the barn owl. Exp Brain Res *92:* 194–208.
Matsuda, K., M. Uemura, M. Kume, R. Matsushima, N. Mizuno (1978) Topographical representation of masticatory muscles in the motor trigeminal nucleus in the rabbit: An HRP study. Neurosci Lett *8:* 1–4.
Mizuno, N., K. Matsuda, N. Iwahori, M. Uemura-Sumi, M. Kume, R. Matsushima (1981) Representation of the masticatory muscles in the motor trigeminal nucleus of the macaque monkey. Neurosci Lett *21:* 19–22.
Mogoseanu, D., A.D. Smith, J.P. Bolam (1994) Monosynaptic innervation of facial motoneurones by neurones of the parvicellular reticular formation. Exp Brain Res *101:* 427–438.
Nakamura, Y., N. Katakura (1995) Generation of masticatory rhythm in the brainstem. Neurosci Res *23:* 1–19.
Nishikawa, K.C., C.W. Anderson, S.M. Deban, J.C. O'Reilly (1992) The evolution of neural circuits controlling feeding behavior in frogs. Brain Behav Evol *40:* 125–140.
Nottebohm, F. (1993) The search for neural mechanisms that define the sensitive period for song learning in birds. Neth J Zool *43:* 193–234.
Okubo, J., S. Kitamura, K. Ogata, A. Sakai (1987) Localization of rabbit laryngeal motoneurons in the nucleus ambiguus. Exp Neurol *96:* 528–539.
Reiner, A., H.J. Karten (1982) Laminar distribution of the cells of origin of the descending tectofugal pathways in the pigeon *(Columba livia)*. J Comp Neurol *204:* 165–187.
Reinke, H., J.M. Wild (1997) Distribution and connections of inspiratory premotor neurons in the brainstem of the pigeon *(Columba livia)*. J Comp Neurol *379:* 347–362.
Rokx, J.T.M., P.J.W. Jüch, J.D. van Willigen (1986) Arrangement and connections of mesencephalic trigeminal neurons in the rat. Acta Anat *127:* 7–15.
Satomi, H., T. Ninomiya, C. Abe, T. Takahashi (1985) Are cat lingual muscles dually innervated by the hypoglossal nerve and the facial nerve? Exp Neurol *87:* 578–582.
Silver, R., P. Witkovsky (1973) Functional characteristics of single units in the spinal trigeminal nucleus of the pigeon. Brain Behav Evol *8:* 287–303.
Steeves, J.D., G.N. Sholomenko, D.M.S. Webster (1987) Stimulation of the pontomedullary reticular formation initiates locomotion in decerebrate birds. Brain Res *401:* 205–212.
Striedter, G.F. (1994) The vocal control pathways in budgerigars differ from those in songbirds. J Comp Neurol *343:* 35–56.
Suthers, R.A., F. Goller, R.S. Hartley (1996) Motor stereotypy and diversity in songs of mimic thrushes. J Neurobiol *30:* 231–245.
Tellegen, A.J. (1997) The Organization of Medullary Reticular Premotor Areas in the Mallard (*Anas platyrhynchos* L.). Control of Head and Jaw Movements; PhD thesis, Leiden University.
Tellegen, A.J., J.L. Dubbeldam (1994) Location of premotor neurons of the motor nuclei innervating craniocervical muscles in the mallard (*Anas platyrhynchos* L.). Eur J Morphol *32:* 138–141.
Tellegen, A.J., J.L. Dubbeldam (1996) Do craniocervical and jaw motor nuclei receive input from the same population of reticular premotor neurons? A double labeling tracing study in the mallard *(Anas platyrhynchos)*. Neurosci Lett *209:* 77–80.
Tellegen, A.J., J.L. Dubbeldam (1999) The location of reticular premotor areas of a motor center innervating craniocervical muscles in the mallard (*Anas platyrhynchos* L.). J Comp Neurol (in press).
Tellegen, A.J., A.M. Karssen, T.M. Rietveld, J.L. Dubbeldam (1998) A crossed projection from the optic tectum to craniocervical premotor neurons in the brainstem reticular formation. An anterograde and retrograde tracing study in the mallard (*Anas platyrhynchos* L.). Eur J Morphol *36* (in press).
Tomioka, S., N. Nakajo, M. Takata (1997) Excitation of hypoglossal motoneurons responsible for tongue protrusions is associated with palatally induced jaw-closing reflex. Neuroscience *78:* 603–612.
Travers, J.B., N. Montgomery, J. Sheridan (1995) Transneuronal labeling in hamster brainstem following lingual injections with herpes simplex virus-1. Neuroscience *68:* 1277–1293.
Van Gennip, E.M.S.J. (1988) A Functional Morphological Study of the Feeding System in Pigeons (*Columba livia* L.). Behavioral Flexibility and Morphological Plasticity; PhD thesis, Leiden University.
Van Gennip, E.M.S.J., H. Berkhoudt (1992) Skull mechanics in the pigeon, *Columba livia*. A three-dimensional kinetic model. J Morphol *213:* 197–224.
Van Willigen, J.D., P.J.W. Jüch, C.M. Ballintijn, M.L. Broekhuijsen (1986) A hierarchy of neural control of mastication in the rat. Neuroscience *19:* 447–455.
Vicario, D.S. (1991) Organization of the zebra finch song control system. II. Functional organization of outputs from nucleus robustus archistriatalis. J Comp Neurol *309:* 486–494.
Vicario, D.S. (1993) A new brain stem pathway for vocal control in the zebra finch song system. Neuroreport *4:* 983–986.
Vu, E.T., M.E. Mazurek, Y.C. Kuo (1994) Identification of a forebrain motor programming network for the learned song of zebra finches. J Neurosci *14:* 6924–6934.
Westneat, M.W., J.H. Long, W. Hoese, S. Nowicki (1993) Kinematics of birdsong: Functional correlation of cranial movements and acoustic features in sparrows. J Exp Biol *182:* 147–171.
Wild, J.M. (1981) Identification and localization of the motor nuclei and sensory projections of the glossopharyngeal, vagus and hypoglossal nerves of the cockatoo *(Cacatua roseicapilla)* Cacatuidae. J Comp Neurol *203:* 351–377.
Wild, J.M. (1990) Peripheral and central terminations of hypoglossal afferents innervating lingual tactile mechanoreceptive complexes in Fringillidae. J Comp Neurol *298:* 157–171.
Wild, J.M. (1993a) The avian nucleus retroambigualis: A nucleus for breathing, singing and calling. Brain Res *606:* 319–324.

Wild, J.M. (1993b) Descending projections of the songbird nucleus robustus archistriatalis. J Comp Neurol *338:* 225–241.
Wild, J.M. (1997) Functional anatomy of neural pathways contributing to the control of song production in birds. Eur J Morphol *35:* 303–325.
Wild, J.M., H.P. Zeigler (1980) Central representation and somatotopic organization of the jaw muscles within the facial and trigeminal nuclei of the pigeon. J Comp Neurol *192:* 175–201.
Zhang, S.P., R. Bandler, P.J. Davies (1995) Brain stem integration of vocalization: Role of the nucleus retroambigualis. J Neurophysiol *74:* 2500–2512.
Zijlstra, C., J.L. Dubbeldam (1994) Organization of the motor innervation of craniocervical muscles in the mallard, *Anas platyrhynchos* L. J Brain Res *35:* 425–440.
Zweers, G. (1982a) The feeding system of the pigeon (*Columba livia* L.). Adv Anat Embryol Cell Biol *73:* 1–108.
Zweers, G.A. (1982b) Pecking in the pigeon (*Columba livia* L.). Behaviour *81:* 173–230.
Zweers, G.A., A.F.Ch. Gerritsen, P.J. van Kranenburg-Voogd (1977) Mechanics of feeding of the mallard (*Anas platyrhynchos* L.; Aves, Anseriformes). Contrib Vertebr Evol. Basel, Karger, vol 3, pp 1–108.
Zweers, G.A., H.C. van Pelt, A. Beckers (1981) Morphology and mechanics of the larynx of the pigeon (*Columba livia* L.): A drill-chuck system (Aves). Zoomorphology *99:* 37–69.